CRYSTAL CLEAR

CRYSTAL CLEAR SELF-DEFRAGMENTATION PROGRAM
A MARDUKITE SYSTEMOLOGY PUBLICATION

CRYSTAL CLEAR

THE SELF-ACTUALIZATION MANUAL & GUIDE TO TOTAL AWARENESS

BY JOSHUA FREE

FOREWORD BY KYRA KAOS

© 2020, JOSHUA FREE

ISBN : 978-0-578-61429-8

No part of this publication may be reproduced in any form or by any means, electronic or mechanical, including photocopying, recording, or any information storage or retrieval system, without permission from the publisher. The information in this book is not intended to substitute any medical treatment or professional advice.

A MARDUKITE SYSTEMOLOGY PUBLICATION

Systemology Air Command Reports #9OCT19—#7NOV19
Mardukite Research Library Catalogue No. "Liber-2B"

Revised Second Printing—April 2020
ALSO AVAILABLE IN HARDCOVER

Published in conjunction with the
NexGen Systemological Society (*NSS*)
& International School of Systemology

Cover Graphics and Systemology Logos by Kyra Kaos

All interior illustrations are photographic enhancements of
Joshua Free's hand-drawn chart/overhead lecture diagrams.

CRYSTAL CLEAR SELF-DEFRAGMENTATION PROGRAM

Take control of your destiny and chart the first steps toward your own spiritual evolution. Realize new potentials for the Human Condition in Self-Honesty. Reclaim the freedom of the spirit with a manual of new techniques and teachings so profound that its effectiveness has been deemed *Crystal Clear*.

NexGen Systemology is grounded on foundations extending back on the timeline of human history to the original inception of current "systems" inherent in our modern world. These records—mostly maintained on *cuneiform tablets* from Mesopotamia—are the *"Arcane Tablets"* within *Systemology* literature and its predecessor, *Mardukite Zuism*, described in a former volume: *"Tablets of Destiny: Using Ancient Wisdom to Unlock Human Potential"* by Joshua Free.

Although many interpretations, analytical philosophies and academic approaches have already applied their treatments to remnants of the Ancient Mystery School (to the extent of their own paradigms), the fact remains that the deepest meanings *realized* of the *"Arcane Tablets"*—from thousands and thousands of years ago—is only *now* being *actualized* toward a full application of a "futurist spiritual technology" for the first time in modern history. These uncharted territories have *always* been there for us. And as we travel this *Pathway* we are constantly reminded how many elements on this journey are *all too familiar*—and *realize* that we *have* actually charted this adventure once before. But, *of course we have*—for this is, and always has been, the *Adventure of Self*.

When a *Seeker* attains actualized *Awareness* outside of physical existence, therein alone lies the true personality of the *Spirit*, the individuated "I" that is *Self*. This is the state of *Self* free of worldly *fragmentation*. The *Seeker* has extended the reach of *Awareness* and the ability (and responsibility) of *Self-direction* from the point of WILL as *Cause* in the Universe, rising above lower planes of *Effect* and *Desire*.

In Joshua Free's "Crystal Clear Self-Defragmentation Course" a *Seeker* may learn to free the *Self* from lower levels of personal *fragmentation* and *programming*; to apply *Will-Intention* to the fullest extent from outside lower "systems"; and live without regard of former *fragmentation* and erroneous *programming* concerning personal ability and responsibility accumulated during this lifetime. The *way out* is always the way *through*; not avoidance, neglect or indifference—which only mark a path to ignorance.

Every technique and *Self-Processing* method contained within "*Crystal Clear*" targets personal management of emotion, memory and thought, assisting *Seekers* to regain *Self-control* of these systems. This is practiced repeatedly, because we are dealing with *conditioning*. And we are not using *Systemology* to condition ourselves into some false sense of grandeur. What we *are* doing is systematically removing *fragmentation* that undeniably reveals the "*Unconditioned Self*"—the very *Alpha* state of our true spiritual *Beingness*. There is no higher purpose or demonstration of the "Great Work" available.

Virtually every other cultural mythology, pantheist tradition and spiritual paradigm has attempted *to be* the route toward *Ascension*—and perhaps within each parameter of the System across time and geography, those methods and ideals were executed to some degree of *effectiveness*. But, we are in the 21st Century now, beginning a decade of 2020's—and it is time we brought some *clear 20/20 vision* to the "I" that is *Self*, and actualize a new *realization* of the Human Condition that unfolds the latent *genetic and spiritual* potential that is already there and just waiting to advance us toward attaining the state of *Homo Novus* in *this* lifetime!

This is *the most important* thing an individual can be *doing right now* for themselves, for family, for society, for all *Life* on the planet—and to every sphere of influence and existence that we can extend our reach in *Self-Honesty* to *infinity*.

CRYSTAL CLEAR: SELF-DEFRAGMENTATION COURSE OUTLINE

Editor's Note 9
Introduction 11

UNIT 1 : SELF-EVALUATION 15

The Systemology "Beta-Awareness Test" (BAT) • Self-Evaluation Questionnaire • Self-Analysis • Self-Evaluation Test (SET) Worksheet • Advanced Self-Analysis • Personal Evaluation • Self-Evaluation Test (SET) Analysis Summary • Self-Evaluation Test (SET) Analysis Graph • The "Standard Model" of NexGen Systemology • The "Emotimeter" and "Awareness Scale" • Understanding the "Emotimeter" • Actualization of Basic Needs • Developing Scales to Chart Systems • The Systemology of Spiritual Immortality • Scales and Degrees on a Continuity Model • Using the "Beta-Awareness Scale"

UNIT 2 : SELF-PROCESSING 67

Systemology "Self-Processing" (Defined) • The Purpose of Processing Toward Self-Honesty • Operating "Self-Processing" Toward Self-Honesty • Managing "Self-Processing" Sessions • An Introduction to Kenostic Processing —"AR-SP-2" [Extended Course]

UNIT 3 : SELF-AWARENESS 97

Raising Personal Levels of Awareness • First Steps with "Self-Processing" • Language and Communication of "Self-Processing" • The Spiritual Side of Systemology Processing • The Physical Side of Systemology Processing • Self-Processing Attention Patterns • Defragmenting "Self-Consciousness" • The Game of "Self-Consciousness" • Managing Conscious Awareness • Self Awareness of Self-Consciousness [Extended Course]

UNIT 4 : SELF-DISCOVERY 163

The Discovery of Self-Consciousness • Self-Realization of the "Light-Body" • Actualization of the "Light-Body" • Systematization of Self-Consciousness • Considerations: Thought, Will & Effort • Rehabilitation of Self-Consciousness • Assuming Alpha Command of Self • Subjective Processing • Objective Processing • Self-Discovery of Human Ability [Extended Course] • Systemology Process "SP-2B-8" Described

UNIT 5 : SELF-DIRECTION 241

"I Will—Therefore I-AM" • Self-Directing Will & Intention • Returning the Reign of Self-Direction • Self-Directing Beyond Guilt and Blame • Self-Directing Cause for Emotional Effect • Objective Processing (Self-Directing Will, Part A & B) • Objective Processing "Part-C" • Self-Directing Will in Self-Honesty • Systemology Process "SP-2B-8A" • Self-Directing Mental Imagery • All Proceeds From and Through Will

UNIT 6 : SELF-DETERMINATION 293

The Route to Self-Determinism and Beyond • Sympathy Imprints and "Excuses" • Managing Sympathy and Low-Energy • Willingness & Self-Determination • Consideration and Willingness • The Reunion of Awareness and Self • Willing "To Be" A Self-Determined Spirit • Subjective Willingness-Processing • Self-Actualizing the Will "To Be" • Emoticurves: Mastering the Pendulum Swing • Death and Initiation: Recycling the Artificial • A Funeral for Beta-Personality

Appendix I 335
Appendix II 336
Appendix III 337
Glossary (ver. 2.2B) 339
Further Reading.......................... 363

∞

EDITOR'S NOTE

"The Self does not actualize Awareness past a point not understood."
—*Tablets of Destiny*

While preparing this book for publication, the editors have made every effort to present the material in a straightforward manner—using clear, easy to read and understand language.

Wherever appropriate, ambiguous and archaic terms are defined as numbered footnotes and in the glossary.

A clear understanding of this material is critical for effective comprehension and personal benefit from *Mardukite Zuism* applied philosophies and *NexGen Systemology* spiritual technology.

The *Seeker* should be especially certain not to simply "read through" this book without attaining proper comprehension as "knowledge." Even when the information continues to be "interesting"—if at any point you find yourself feeling lost or confused while reading, trace your steps back. Return to the point of misunderstanding—and go through it again.

It is expected that a *Seeker* will work through this cycle of materials multiple times to achieve optimum results.

And *now* responsibility for this power and its actualization is passed on to you, the *Seeker*.

Take nothing within this book on faith. Apply the information directly to your life. Decide for yourself.

—NexGen Systemological Society (NSS)

∞

*A gift to homo sapiens at
the birth of homo novus*

—J.F.

To The Chamberlains:

Your contributions without measure,
Your company a source of pleasure.

The amount of effort you have given,
without any rhyme or reason,

is incomparable
to the work.

A thing
does not
exist

to equal your worth.
Except, these books.

—K.K.

CRYSTAL CLEAR : MOVING FORWARD
an Introduction by Kyra Kaos

SALUTATIONS SEEKERS! WELCOME!

Here on your journey, you have come across something unexpectedly—something truly incredible. For this *book* is not *just a book*; it is not one of those *"read-it-and-be-done-with-it"* type of books. (Not that many books by Joshua Free are.) This *book*—now in your hands—is literally the most effective *tool* we can provide for you to reclaim the power to actually *change your life!*

If you have come across the *Crystal Clear Self-Defragmentation Course* as a result of following the previous laid pathway of works of Joshua Free, especially the immediate precursor, *"The Tablets of Destiny"*—often referred to in this present volume as *Liber-One*—then *congratulations* on making it this far! For those of you that have not read *Liber-One*, don't despair—but I would strongly encourage you to do so, because it introduces, for the first time in modern history, many paramount definitions and key concepts of *Systemology* found throughout the present volume. As such, a newcomer should pay particular attention to the definitions provided in the footnotes and glossary of *Crystal Clear*.

For those who have highly anticipated this follow-up to *Liber-One*: That's right, it's finally here! The cosmic *crystal clarity* has arrived! Now that the wisdom of the *Arcane Tablets* have been effectively harvested, cultivated and synthesized into a working *systemology*, all that remains is for the *Seeker* to actually read, learn and perform the workable techniques of *Crystal Clear* that are designed to tap into the latent potential entombed within every human on the planet. Yep, there is still some work *you* have to do—but, thanks to the "New Thought Division" of the "Mardukite Research Organization" called *"NexGen Systemology,"* you don't have to do much *digging*; as it has all been laid out *crystal clear* before you.

This work defines an *Alpha* point of *Beingness* where you and every Human, since the inception of the species, has sought a return to. This has been no small feat accomplished by Joshua Free and those of us assisting in the work at the Mardukite Offices. And these efforts have not been discounted—so no small "Thank You" can be extended to all of you that have been sticking with us throughout the ages.

> "As above, so below; on earth, as it is in heaven."
> —The Tablets of Destiny

We each chose—that's right, *even you*—to be an "existence" *as* these bodies, these *genetic vehicles*[1] that we control to experience an existence on this planet. *I am* and *you are* an *Alpha Spirit*[2] controlling a body; not the physical "beta" body itself. This physical body is only a solidification of physical energy and physical matter—something you have learned to manipulate at *Will* with practice. *NexGen Systemology* extends this "practice" to *all* "systems" that may be governed by *Will*; which is *all systems*.

As fragments of *Spiritual Identity* in a physical body, we are forced to observe and interact with *others* following the *course* of their own path in this *beta existence*.[3] It is here that

---

1 **genetic-vehicle** : a physical *Life*-form; the physical (*beta*) body that is animated/controlled by the (*Alpha*) *Spirit* using a continuous *Lifeline* (ZU); a physical (*beta*) organic receptacle and catalyst for the (*Alpha*) *Self* to operate "causes" and experience "effects" within the *Physical Universe*.
2 **alpha-spirit** : a "spiritual" *Life*-form; the "true" *Self* or I-AM; the spiritual (*Alpha*) *Self* that is animating/controlling the (*beta*) physical body or *genetic-vehicle* using a continuous *Lifeline* of spiritual energy (ZU); an individual spiritual (*Alpha*) entity possessing no physical mass or measurable waveform (motion) in the Physical Universe (KI) as itself, so it animates/manages the (*beta*) physical body or *genetic-vehicle* as a catalyst to experience *Self-determined* causality (in effect) within the *Physical Universe*.
3 **beta-existence** : all manifestation in the "Physical Universe" (KI); the "Physical" state of existence consisting of vibrations of physical energy and physical matter moving through physical space and

most people experience little more than a reality of "work/eat/sleep/repeat" and the nature of this beast: a snake-eating-it's-own-tail. All while you are also trying to figure out what you are doing on this flying ball of mud and rock through space. By the time you get to an end point, all you have really done is work/eat/sleep/repeat. You did not personally manage time and space to figure out *you*, who *you* really are—and chances are in the next life, you will return to a cycle of "work/eat/sleep/repeat" never knowing otherwise.

The *Pathway to Self-Honesty* requires us to break this cycle—no, I don't mean "quit your day job"—and learn what it means for *beta-Awareness*[4] to ascend to true *Knowingness*, then continue on your journey like you were meant to, but with a greater certainty than when you first arrived here. True progress on what we call: *"The Pathway to Self-Honesty"* is accomplished through one thing alone: remembering who *you* are. It is important, above all else in this *beta-existence*, for us to manage our time and space toward this progress.

In *"Tablets of Destiny"* (*Liber-One*) you were granted unique access to a "bridge" from the *"Mardukite Core"* to *"Mardukite Systemology"*—developed as a result of many years of research and experimental dedication from our faithful members. *Liber-One* introduces the main premise of this work, its concepts and terminology—all of which lead a *Seeker* in the direction of greater personal management and *Self-control*; all of which may *now* be fully actualized by *any one* with the

---

   experienced as "time"; the conditions of *Awareness* for the *Alpha-Spirit* (*Self*) as a physical organic *Lifeform* or "*genetic vehicle*" in which it experiences causality in the *Physical Universe*.
4 **beta-awareness** : all consciousness activity ("*Awareness*") in the "Physical Universe" (KI) or else *beta-existence*; *Awareness* within the range of the *genetic-body*, including material thoughts, emotional responses and physical motors; personal *Awareness* of physical energy and physical matter moving through physical space and experienced as "time"; the *Awareness* held by *Self* that is restricted to a physical organic *Lifeform* or "*genetic vehicle*" in which it experiences causality in the *Physical Universe*.

dedication to simply apply the instruction and *Self-Processing* of the *Crystal Clear Self-Defragmentation Course* to their daily lives. This amazing *book* is a priceless gem for true *Seekers*, specially designed in view of the fact that most readers are without direct access to the underground activities of the *Mardukite Office*, or course-work provided by an officially appropriated branch of the *NexGen Systemology Society*.

CRYSTAL CLEAR is The Book *about* YOU!
It demonstrates the power to register your own *Destiny* on the proverbial *Tablets of Life*...

Our present journey begins and ends with an emphasis on *Self*. At the start, and at each step along the way, you are given the resources to properly evaluate your *Self* on the *Pathway*—and the only one you will compare your *Self* to <u>is</u> your *Self*, or a former realization of your *Self*. Everything you need in order to know "where you are" on the *Pathway*—and in any space and at any time is *here*. The power inherent in a *crystal clear* realization of *Self* is already right there within you. Even if you didn't know you had it, *Crystal Clear* will most certainly shine the light on where it is hidden.

Enjoy your journey, *Seekers*.
For you are further than you think.

From the Pathway,

—Kyra Kaos
Halloween, 2019

UNIT ONE
SELF EVALUATION

THE "BETA-AWARENESS TEST" ("BAT")

PURPOSE:

The *"Beta Awareness Test" (BAT)* was developed for *Mardukite Systemology* to determine a "basic" or "average" state of personal *Awareness*—specifically:

Beta-Awareness maintained by an individual *Alpha-Spirit* in its management of a *genetic-vehicle* to experience *beta-existence*.

A personal assessment of each "test" statement is rated and scored with an "assessment scale"[5] based on the systematic relationship demonstrated by the *NexGen Systemology Standard Model*,[6] first introduced in *"Tablets of Destiny: Using Ancient Wisdom to Unlock Human Potential."* Additional *psychometric evaluation*[7] of this assessment allows an individual[8] to "chart" personal *Awareness (ZU)*[9] according to "degrees" graphically

---

5 **assessment scale** : an official assignment of graded/gradient numeric values.
6 **standard model** : a fundamental *structure* or symbolic construct used to evaluate a complete *set* in *continuity* relative to itself and variable to all other *dynamic systems* as graphed or calculated by *logic*; in *NexGen Systemology*—a *"monistic continuity model"* demonstrating *total system* interconnectivity "above" and "below" observation of any apparent *parameters*; the *ZU-line* represented as a singular vertical (*y*-axis) waveform in space without charting any specific movement across a dimensional time-graph *x*-axis.
7 **psychometric evaluation** : the relative measurement of personal ability, mental (psychological/thought) faculties, and effective processing of information and external stimulus data; a scale used in "applied psychology" to evaluate and predict human behavior.
8 **individual** : a person, human entity or *Seeker* is sometimes referred to as an "individual" within this text.
9 **ZU** : the ancient cuneiform sign designating an archaic verb—*"to know,"* *"knowingness"* or *"awareness"*; the active energy/matter of the "Spiritual Universe" (AN) that is experienced as *Lifeforce* or *consciousness* for entities existing in the "Physical Universe" (KI); *"Spiritual Life Energy"*; the spiritual energy present in the WILL of the actualized *Alpha-Spirit* in the "Spiritual Universe" (AN), which imbues its *Awareness* into the Physical Universe (KI), animating or controlling *Life* for its experience of *beta-existence* along an *Identity-*

demonstrated by the *ZU-line*[10] (combined with the *Standard Model*) using the "average" *BAT* score.

INSTRUCTIONS:

Being as *honest* as you are able, complete each of the following statements using the *Assessment Scale* provided. With a pencil, write down the "number" (*numeric value*) from the *Assessment Scale* next to each statement. [Mark a "2" if you honestly "don't know" or are really uncertain.] Don't be alarmed if you find many of the *assessment values* to be lower at first. The basic goal of *NexGen Systemology* is to increase certainty and ability for all of the qualities listed in the questionnaire.

You will assess the same *"Basic Awareness Test" (BAT)* several times while on the *Pathway to Self-Honesty* to chart your progress using *Mardukite Systemology Tech*. You may prefer to mark your answers directly onto the worksheet provided for evaluation. Use *"Self-Evaluation Test (SET) Worksheet #1"* the first time you read this far in the present book. [Additional copies of this worksheet are included in this volume for use as directed.] Only include *numeric values* when testing to make certain you are not inadvertently performing "self-hypnosis" or *autosuggestion*[*] using inadequate affirmations during *Self-Evaluation*.[11]

---

continuum called a *ZU-line*.

10 **ZU-line** : a spectrum of *Spiritual Life Energy*; an energetic channel of *Identity-continuum* connecting the *Awareness (ZU)* of an *Alpha-Spirit* with *"Infinity"*; a *Life-line* on which *Awareness (ZU)* extends from the direction of the "Spiritual Universe" (AN) as its *alpha state* through an entire possible range of activity in its *beta state*, experienced as a *genetic-entity* occupying the *Physical Universe (KI)*.

* If an individual so desired, these statements *could be* reworked as positive affirmations, by inserting the most idealistic statement of "always" in each. However, such should not be performed as a substitution for true *Self-Evaluation*. Doing so could greatly hinder *actualized* progress on the *Pathway to Self-Honesty*.

11 **Self-evaluation** : see *"psychometric evaluation"* and *"BAT."*

THE NEXGEN SYSTEMOLOGY
SELF-EVALUATION QUESTIONNAIRE

A-1 ___ = I am ___ successful at physical activities.
B-1 ___ = I ___ maintain control of my emotions.
C-1 ___ = I am ___ certain of what I know.
D-1 ___ = I am ___ in control of my reality.
E-1 ___ = I ___ make choices easily.
F-1 ___ = I am ___ understood by others.
G-1 ___ = I ___ find my life experience to be pleasurable.
H-1 ___ = I am ___ satisfied with my accomplishments.

ASSESSMENT SCALE
- 0 = Never
- 1 = Rarely
- 1.5 = Seldom
- 2 = Sometimes
- 2.5 = Usually
- 3 = Very Often
- 3.5 = Consistently
- 4 = Always

I-1 ___ = I ___ solve my problems with goals and life successfully.
J-1 ___ = I ___ appreciate beauty in life, nature and the universe.
K-1 ___ = I ___ express myself warmly and cheerfully.
L-1 ___ = I ___ provide my highest quality of work.
M-1 ___ = I ___ take responsibility easily.
N-1 ___ = I am ___ respectful of traditions and opinions of others.
O-1 ___ = I am ___ sensitive to needs of others without succumbing to sympathetic emotion.
P-1 ___ = I am ___ cooperative and successful in a group.
Q-1 ___ = I ___ know and trust myself, even in times of stress.
A-2 ___ = I ___ maintain my physical health and well-being.
B-2 ___ = I ___ maintain my composure when others are excited.
C-2 ___ = I ___ freely determine where to set my attentions and interests.

D-2 ___ = I am ___ the cause of my thoughts, ideas and emotions.
E-2 ___ = I ___ promptly make good decisions.
F-2 ___ = I ___ find it easy to communicate my thoughts and feelings to others.
G-2 ___ = I am ___ in control of my life.
H-2 ___ = I am ___ enthusiastic about my life and treat it as an adventure.
I-2 ___ = I ___ achieve my goals and objectives.
J-2 ___ = I ___ take quality time for myself, quietly, to think.
K-2 ___ = I ___ express my thoughts, feelings and abilities easily, openly and honestly.
L-2 ___ = I ___ consider effects of my decisions on others.
M-2 ___ = I ___ take responsibility for my emotional state.
N-2 ___ = I am ___ tolerant to beliefs of others.
O-2 ___ = I am ___ empathic to emotions of others without taking them on as my own.
P-2 ___ = I ___ set reasonable standards for myself and others.
Q-2 ___ = I ___ keep promises and repay debts.
A-3 ___ = I am ___ able to maintain good health, even when others in my environment are ill.
B-3 ___ = I ___ maintain my emotions and reactions under stress.
C-3 ___ = I can ___ change my mind, decisions or beliefs easily.
D-3 ___ = I am ___ able to manage my environment.
E-3 ___ = I ___ act to resolve a problem when I see it.
F-3 ___ = I ___ understand anything I decide to learn.
G-3 ___ = I am ___ in healthy and clean physical condition.
H-3 ___ = I am ___ content with my status and progress in life.
I-3 ___ = I ___ enjoy thinking about my future.
J-3 ___ = I am ___ able to enjoy my solitude.

K-3 ___ = I ___ exchange, express and receive ideas freely, openly and easily.
L-3 ___ = I ___ gain trust and support from others with my enthusiasm.
M-3 ___ = I ___ take responsibility for my thoughts and actions.
N-3 ___ = I ___ conduct myself without resentment.
O-3 ___ = I am ___ able to assist someone in need without becoming emotional.
P-3 ___ = I ___ consider what happens to others.
Q-3 ___ = I ___ meet my obligations and fulfill promises.

ASSESSMENT SCALE	
0	= Never
1	= Rarely
1.5	= Seldom
2	= Sometimes
2.5	= Usually
3	= Very Often
3.5	= Consistently
4	= Always

A-4 ___ = I ___ keep my possessions neat and orderly
B-4 ___ = I ___ avoid feeling gloomy and depressed.
C-4 ___ = I ___ think clearly in intense, emotional or stressful situations.
D-4 ___ = I ___ affect those in my environment with my thoughts and emotions.
E-4 ___ = I ___ act to resolve needless suffering when I see it.
F-4 ___ = I ___ know when to speak—and when to remain silent
G-4 ___ = I ___ calculate risks effectively before I act.
H-4 ___ = I am ___ able to adapt to any situation.
I-4 ___ = I ___ set future goals and achieve them.
J-4 ___ = I am ___ able to clearly *imagine* images/sounds/smells/any sensation that I have *not* experienced.
K-4 ___ = I am ___ able to express my affections.
L-4 ___ = I ___ can find something to be enthusiastic about.
M-4 ___ = I ___ maintain control of my time management.
N-4 ___ = I am ___ patient in my interactions with others.

O-4 ___ = I ___ can sense when others silently react or (mis)understand my communications.
P-4 ___ = I am ___ sought by others for advice and counsel.
Q-4 ___ = I ___ determine the truth of matters I am interested in.
A-5 ___ = I ___ appreciate and take care of my property.
B-5 ___ = I ___ put emotional memories aside when solving present problems or working toward future goals.
C-5 ___ = I ___ can recall past events clearly and vividly.
D-5 ___ = I ___ make every effort to fully utilize and advance my knowledge and ability.
E-5 ___ = I ___ continue a task to completion, even if I make mistakes.
F-5 ___ = I ___ find out the information when I do not fully understand something.
G-5 ___ = I am ___ able to clearly *recall* images/sounds/smells/any sensation that I *have* experienced in the past.
H-5 ___ = I ___ can fully focus my concentration on the present task at hand.
I-5 ___ = I am ___ courageous, upfront and forthcoming in my behavior.
J-5 ___ = I ___ occupy my time and efforts creating, building or designing something.
K-5 ___ = I ___ share my thoughts and ideas with others.
L-5 ___ = I am ___ considered a cheerful person by others.
M-5 ___ = I ___ can cause change in my environment at will.
N-5 ___ = I ___ can receive criticism from others without exhibiting an emotional reaction.
O-5 ___ = I ___ assist others to understand when my communication is misunderstood.
P-5 ___ = I am ___ considered dependable by others.
Q-5 ___ = I am ___ truthful and honest to myself and others.

A-6 ___ = I ___ take care of, maintain and treat my physical environment well.
B-6 ___ = I ___ *sense* an emotional state before I *feel* it.
C-6 ___ = I ___ think positive thoughts about myself and others.
D-6 ___ = I ___ think and act systematically.
E-6 ___ = I ___ have close friends I can confide in.
F-6 ___ = I ___ seek to improve an ability or increase a skill.
G-6 ___ = I am ___ able to discard old memorabilia freely.
H-6 ___ = I ___ avoid borrowing money or using credit cards.
I-6 ___ = I ___ act to protect and maintain our planet and ecosystem on earth.
J-6 ___ = I am ___ able to create new things and develop new ideas.
K-6 ___ = I ___ can motivate others to action.
L-6 ___ = I am ___ considered charismatic by others.
M-6 ___ = I ___ control the urge for excess or to overindulge.
N-6 ___ = I ___ restrain my display of anger toward others.
O-6 ___ = I am ___ welcoming to the company of children and animals.
P-6 ___ = I ___ keep my office/study neat and organized.
Q-6 ___ = I am ___ considered trustworthy by others.

ASSESSMENT SCALE	
0	= Never
1	= Rarely
1.5	= Seldom
2	= Sometimes
2.5	= Usually
3	= Very Often
3.5	= Consistently
4	= Always

SELF-ANALYSIS

This *"Beta-Awareness Test" (BAT)* may be scored and evaluated —*analyzed*—in several ways.

The most <u>basic method</u> of evaluation also carries the most advanced requirement—an experienced *NexGen Systemologist* with sufficient knowledge of the subject and *Awareness Chart* (introduced within this present volume) to plot a score for themselves using just a few key assessments that are known with certainty.

The <u>standard method</u> requires you to write down a *numeric assessment value* for each statement directly onto a *"Self-Evaluation Test (SET) Worksheet"*—as provided. There is a space on the worksheet for all key pieces of information: each *individual answer*; the totals and averages for answered *rows* (such as "A"); and the total and average of a *completed* worksheet.

> <u>STEP #1</u>: Begin with the value for "A-1" and proceed to the space below it for "B-1" and so on. At the end of the questionnaire, you will have completed—and assigned values to—approximately 100 statements. Once these values are all marked on the worksheet, certain observations or patterns may already become apparent[12]— but we make further calculations anyway.

For example: it may become obvious when any particular value—such as "2.0" or "2.5"—appears more frequently than any other, thus somewhat accurately representing the total average itself; or it may be noticed that all values contained within certain rows are significantly higher or lower than those found in other rows—thus somewhat accurately representing particular areas of ability and deficiency reflected in a *Seeker's* current or present state of *Awareness*.

---

12 **apparent** : visibly exposed to sight; evident rather than actual, as presumed by Observation; readily perceived, especially by the senses.

STEP #2: We effectively seek (and accurately use) "averages" for evaluation because specific "qualities" addressed in the questionnaire—as systematized in each *row*—each have a strong tendency to "rise and fall" together. Likewise, the "total average" for a completed worksheet may be used to plot a *Seeker's* base "level" on the *Awareness Chart*, because all of these "qualities" are also systematically interconnected—thus quite accurately representing an overall demonstration of the *Seeker's* personal management of *Reality Experience* in *Self-Honesty*.[13]

TOTALS for a *row* (A-1 + A-2 + A-3...) may be marked on the worksheet at the end of the row. There is also a space for the option of calculating the row AVERAGE. If you have difficulty calculating the AVERAGE for rows, don't worry. Simply add up the TOTAL of each row to write in the "TOTAL" column for that row and skip to *Step #3*. The TOTAL of each row is more important for overall completed test accuracy than an individual row AVERAGE, which you may estimate as needed.

It is true that the *exact* AVERAGE for a row is equal to the TOTAL of its six values *divided* by six—but this is *not* necessary. If all of the values in a row are the same, than the AVERAGE will be the same—or, if a certain value appears most often, you can simply use that value as an AVERAGE estimation for the row.

STEP #3: A *completed test* is quite simple to score and evaluate. Write the TOTAL for the *"SET Worksheet"* by adding up the *column* of values for each individual row TOTAL. In most cases, this provides a three-digit or three-figure value, meaning in the hundreds. Because

---

13 **Self-Honesty** : the *alpha* state of *being* and *knowing*; clear and present total *Awareness* of-and-as *Self*, in its most basic and true proactive expression of itself as *Spirit* or *I-AM*—free of artificial attachments, perceptive filters and other emotionally-reactive or mentally-conditioned programming imposed on the human condition by the systematized physical world.

we have approximately 100 questions or values, the completed test AVERAGE is easily calculated by moving two "decimal places" of the TOTAL. For example: a completed *test total of 220* (*220.*) would be scored as an *average of 2.2* (*2.20*) on the *Awareness Chart*. Or—a completed *test total of 318* would equal an *average of 3.18* on the *Chart*, or *3.2* if we want to "round" for convenience.

Self-Evalutation Test (SET) Worksheet #1

(Use this worksheet the first time.)

AVG.	TOTAL 1-6	6	5	4	3	2	1	
								A
								B
								C
								D
								E
								F
								G
								H
								I
								J
								K
								L
								M
								N
								O
								P
								Q
AVG ___	TOT ___							

Self-Evalutation Test (SET) Worksheet #2

(Use this worksheet after two weeks of daily processing.)

	1	2	3	4	5	6	TOTAL 1-6	AVG.
A								
B								
C								
D								
E								
F								
G								
H								
I								
J								
K								
L								
M								
N								
O								
P								
Q								
TOT ___								
AVG ___								

Self-Evalutation Test (SET) Worksheet #3

(Use this worksheet after two months of processing.)

	A	B	C	D	E	F	G	H	I	J	K	L	M	N	O	P	Q	
1																		
2																		
3																		
4																		
5																		
6																		
TOTAL 1-6																		TOT ___
AVG.																		AVG ___

Self-Evalutation Test (SET) Worksheet #4

(Use this worksheet after six months of processing.)

AVG.	TOTAL 1-6	6	5	4	3	2	1	
								A
								B
								C
								D
								E
								F
								G
								H
								I
								J
								K
								L
								M
								N
								O
								P
								Q
AVG ___	TOT ___							

ADVANCED SELF-ANALYSIS

After average scores are determined, a closer examination of the *"Beta Awareness Test" (BAT)* and completed *"Self-Evaluation Test Worksheet" (SET)* reveals additional data for *advanced* Self-Analysis. It will be noticed that questionnaire statements demonstrate a pattern. There are many reasons *why* our questionnaire is arranged as it is—however, it is much more important for the *Seeker*[14] to understand *how* it is arranged—*after* it is completed the first time.

Letters "A" through "Q" are assigned to statements on the *BAT* questionnaire. There are *six* cycles of these, composing a complete *row* on the *SET* worksheet. Each *row* represents a particular *category*. Each of these categories reflects traits indicative of a specific "personality *quality*" or "*aspect* of *Self* in *beta experience*" that may be evaluated independently—and in systematic relation to the others—with *advanced Self-analysis*.

These categories are distinguished by the following aspects:

SELF-EVALUATION TEST ASPECTS (ADVANCED ANALYSIS)
A = Managing Physical Conditions
B = Managing Emotional Energy
C = Managing Thought Activity
D = Managing External Reality Environment
E = Managing Choice & Decision
F = Managing Communication & Understanding
G = Managing the Past
H = Managing the Present
I = Managing the Future
J = Managing Imagination & Creation
K = Managing Personal Expression (To Others)
L = Managing Personal Magnetism
M = Managing Responsibility & Self-Control

---

14 Seeker : an individual on the *Pathway to Self-Honesty*; a practitioner of *Mardukite Systemology* or *NexGen Systemology Processing* that is working toward *Ascension*.

> N = Managing Personal Tolerance (To Others)
> O = Managing Assistance & Sympathy (To Others)
> P = Managing Interpersonal Organization
> Q = Managing Self-Honesty

The first questionnaire statement for each category (A-1, B-1, &tc.) is a personal assessment of an upper-most expression, consequence or "the *effect*" regarding that aspect. A perfectly executed honest test—in theory—would result in an *average value* for the *row* being equivalent to this answer alone. However, we are aware of many shortcomings inherent in any method of subjective analysis, especially when it comes to ourselves. Therefore, it is critical—for a systematic approach—that we balance or *average* out any *mathematical* evaluation with *"reason"* if we are to effectively plot it on an objective[15] gradient—such as the *Awareness Scale*.

"Blanket statements" are sometimes challenging for an individual to determine for themselves. As a result, we provide the *Seeker* some leeway here by offering five additional *causative statements* to average out a complete assessment. Each subsequent statement within a category, after the first, assesses personal *"cause"*—the degree of *Self-determination* yielding the results or *effects* for that aspect as *beta experience*.

For example: in the first "A-statement" we assess very simply a generalization of how successful we are in physical activities. Each subsequent "A-statement" refers to *Self-directed reasons* why we may be at *that* level of "success" in physical activities—mainly how we treat our body, our physical possessions and manage our immediate physical environment. We should expect *values* to be the same or similar, but since tests and experiments are not written or conducted under absolute (or perfect) conditions, we calculate averages for our purposes—and the results seem to be consistent.

---

15 **objectively** : concerning the "external world" and attempts to observe Reality independent of personal "subjective" factors.

PERSONAL EVALUATION

While developing this test, we discovered that category types evaluated here all share a very interesting attribute in common. They are all the *subject* and *basis* of nearly every *applied* form of "spirituality," "mysticism," or "religious prayer"—essentially any "magical" property or aspect we might wish to *attract* greater "certainty" and "control" of in our *beta existence*. Yet, we soon discover that *all* of these "domains" are inherently *our* "domains" to be responsible for. The actualization[16] of *Self-Honesty* toward directing and "managing" *beta existence* is conditional on—and proportional[17] to—the personal responsibility[18] for *acquiring* "right education" and *applying* "right effort" in each of these categories.

Before commencing with sections and lesson-chapters further demonstrating this "right education" and "right effort," it is important that a *Seeker* is able to clearly evaluate the data from a completed *"SET"* worksheet. It is true that we have used *advanced analysis* in our office to validate[19] and gauge research methods and development of our unique field of *"NexGen Systemology"* and its *Tech*. These same methods are now much more valuable to the *Seeker* for validating and gauging personal progress and individual development on the *Pathway to Self-Honesty*, while using the materials in this present volume. Each time a *"SET"* worksheet is completed, a *Seeker* can mark *averages* for each *row*/category on the following worksheet addendum titled: *"Self-Evaluation Test (SET) Analysis Summary."* This data can be easily plotted on a graph (provided). There is also space allotted next to each "category description" for any additional notes.

---

16 **actualization** : to make actual; to bring into Reality; to realize fully in *Awareness*.
17 **proportional** : having a direct relationship or mutual interaction with.
18 **responsibility** : the *ability* to *respond*; the extent of mobilizing *power* and *understanding* an individual maintains as *Awareness* to enact *change*; the proactive ability to *Self-direct* and make decisions independent of an outside authority.
19 **validation** : the reinforcement of agreements of Reality.

Self-Evaluation Test (SET) Analysis Summary #1

(Use this worksheet addendum the first time.)

	CATEGORY	AVERAGE
A	Physical Condition	
B	Emotional Energy	
C	Thought Activity	
D	External Reality Environment	
E	Choice & Decision Making	
F	Communication & Understanding	
G	Managing the Past	
H	Managing the Present	
I	Managing the Future	
J	Imagination & Creation	
K	Personal Expression to Others	
L	Personal Magnetism	
M	Responsibility & Self-Control	
N	Personal Tolerance to Others	
O	Assistance & Sympathy to Others	
P	Interpersonal Organization	
Q	Managing Self-Honesty	
+	BETA AWARENESS TEST SCORE	

Self-Evaluation Test (SET) Analysis Summary #2

(Use this worksheet addendum after two weeks of daily processing.)

	CATEGORY	AVERAGE
A	Physical Condition	
B	Emotional Energy	
C	Thought Activity	
D	External Reality Environment	
E	Choice & Decision Making	
F	Communication & Understanding	
G	Managing the Past	
H	Managing the Present	
I	Managing the Future	
J	Imagination & Creation	
K	Personal Expression to Others	
L	Personal Magnetism	
M	Responsibility & Self-Control	
N	Personal Tolerance to Others	
O	Assistance & Sympathy to Others	
P	Interpersonal Organization	
Q	Managing Self-Honesty	
‡	BETA AWARENESS TEST SCORE	

Self-Evaluation Test (SET) Analysis Summary #3

(Use this worksheet addendum after two months of processing.)

AVERAGE	CATEGORY	
A	Physical Condition	
B	Emotional Energy	
C	Thought Activity	
D	External Reality Environment	
E	Choice & Decision Making	
F	Communication & Understanding	
G	Managing the Past	
H	Managing the Present	
I	Managing the Future	
J	Imagination & Creation	
K	Personal Expression to Others	
L	Personal Magnetism	
M	Responsibility & Self-Control	
N	Personal Tolerance to Others	
O	Assistance & Sympathy to Others	
P	Interpersonal Organization	
Q	Managing Self-Honesty	
╪	BETA AWARENESS TEST SCORE	

Self-Evaluation Test (SET) Analysis Summary #4

(Use this worksheet addendum after six months of processing.)

	CATEGORY	AVERAGE
A	Physical Condition	
B	Emotional Energy	
C	Thought Activity	
D	External Reality Environment	
E	Choice & Decision Making	
F	Communication & Understanding	
G	Managing the Past	
H	Managing the Present	
I	Managing the Future	
J	Imagination & Creation	
K	Personal Expression to Others	
L	Personal Magnetism	
M	Responsibility & Self-Control	
N	Personal Tolerance to Others	
O	Assistance & Sympathy to Others	
P	Interpersonal Organization	
Q	Managing Self-Honesty	
ǂ	BETA AWARENESS TEST SCORE	

Self-Evaluation Test (SET) Analysis Graph

(*Use different colors to graph changes on this addendum.*)

%	100	85	70	55	40	25	10	5	1	%
Q										Q
P										P
O										O
N										N
M										M
L										L
K										K
J										J
I										I
H										H
G										G
F										F
E										E
D										D
C										C
B										B
A										A
#	4	3.5	3	2.5	2	1.5	1	0.5	0.1	#

Self-Evaluation Test (SET) Analysis Graph

(Use different colors to graph changes on this addendum.)

%	100	85	70	55	40	25	10	5	1	%
Q										Q
P										P
O										O
N										N
M										M
L										L
K										K
J										J
I										I
H										H
G										G
F										F
E										E
D										D
C										C
B										B
A										A
#	4	3.5	3	2.5	2	1.5	1	0.5	0.1	#

Self-Evaluation Test (SET) Analysis Graph—Example Cases

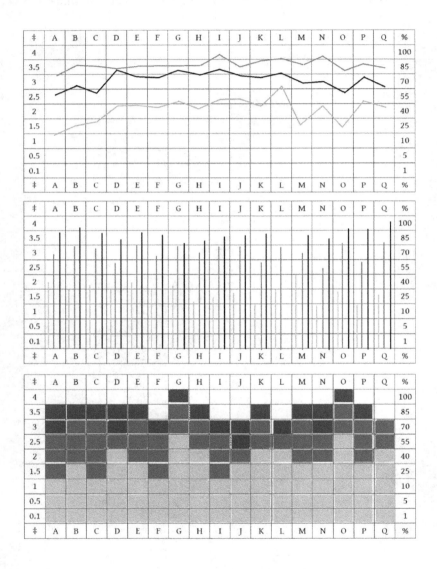

THE "STANDARD MODEL" OF NEXGEN SYSTEMOLOGY

Intensive analysis of ancient *"cuneiform tablets"*[20] from Mesopotamia led to our development of the "Standard Model of Systemology"—accounted for in a previous textbook volume, *Tablets of Destiny: Using Ancient Wisdom to Unlock Human Potential*, also referred to in Mardukite Systemology as *"Liber-One."* Early research and experimentation using *Systemology: The Original Thesis* also led to a discovery and understanding of *"ZU"*—or more specifically, the *"ZU-line"*—and its significance for developing *NexGen Systemology* into an *"applied spiritual philosophy"* that is now communicable and effective.

Our *Pathway to Self-Honesty* in *Mardukite Systemology* is demonstrated with the "Standard Model." The gradient scale by which we chart anything on this *model*—in relation to *Self-Awareness*—is defined as the "ZU-line"—which is a graphic representation of the spiritual energy frequency that we call *Awareness* (or *consciousness*) imbuing all *Life*. This "ZU" energy originates in the "Spiritual Universe" ("AN")[21] as the "I" or "*Self.*" It extends from this *Alpha*[22] state as "spirit" to the "Physical Universe" ("KI")[23] where it maintains a *beta* state by

---

20 **cuneiform** : the oldest extant writing from Mesopotamia; wedge-shaped script inscribed on clay tablets with a reed pen.
21 **AN** : an ancient cuneiform sign designating the *'spiritual zone'*; the *Spiritual Universe*—comprised of spiritual matter and spiritual energy; a direction of motion toward spiritual *Infinity*, away from or superior to the physical (*'KI'*); the spiritual condition of existence providing for our primary *Alpha* state as an individual *Identity* or *I-AM-Self* which interacts and experiences *Awareness* of a *beta* state in the *Physical Universe* (*'KI'*) as *Life*.
22 **alpha** : the first, primary, superior or beginning of some form.
23 **KI** : an ancient cuneiform sign designating the *'physical zone'*; the *Physical Universe*—comprised of physical matter and physical energy in action across space and observed as time; a direction of motion toward material *Continuity*, away from or subordinate to the Spiritual (*'AN'*); the physical condition of existence providing for our *beta* state of *Awareness* experienced (and interacted with) as an individual *Lifeform* from our primary Alpha state of Identity or *I-AM-Self.*

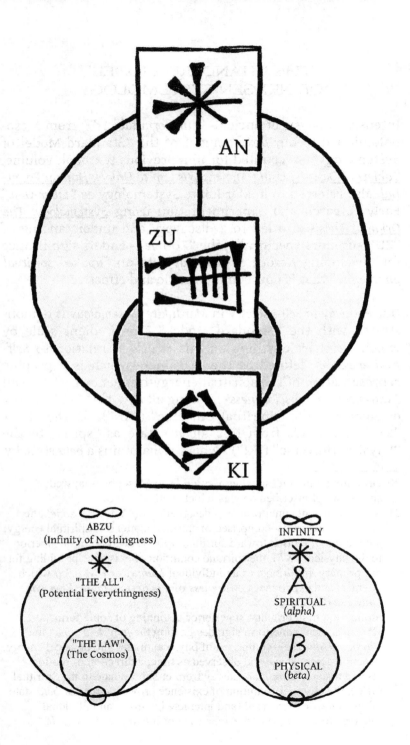

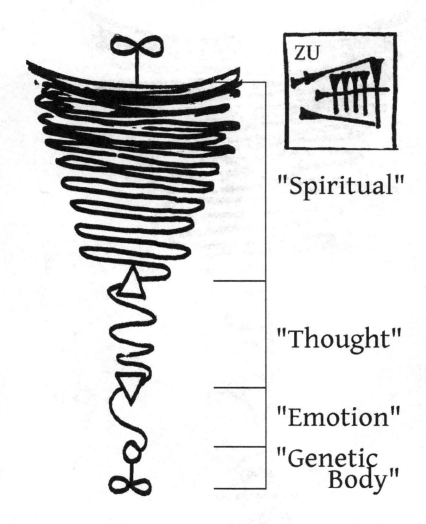

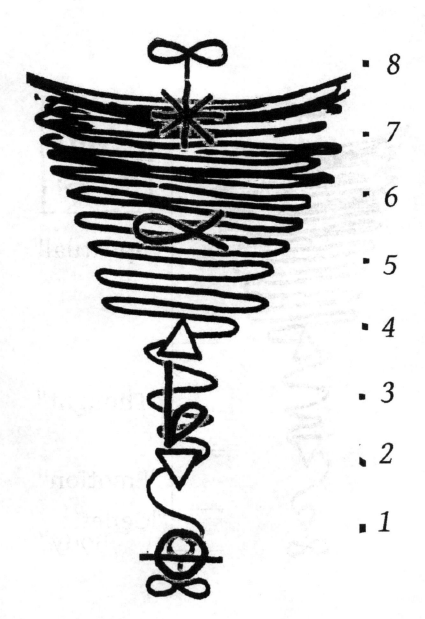

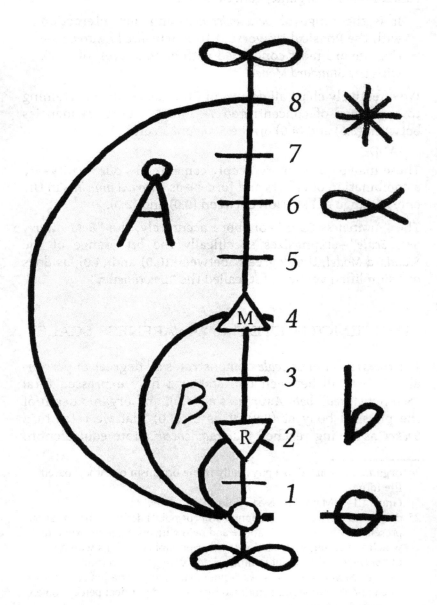

controlling an "organic/genetic body."[24]

It is the range of *beta existence*—and our interaction with the Physical Universe (KI) as *Actualized Awareness*—that we are most concerned with at this level* of work with the *Standard Model*.

We effectively chart all *degrees* of ZU *fragmentation* pertaining to the range of concentrated THOUGHT energy frequencies between (2.0) and (4.0) on the *Standard Model*.

—AND—

These may be even more deeply cemented as our Reality—by a combination of efforts and forces—as *encoded imprints* in the energy range of EMOTION between (0.0) and (2.0).

The *"Awareness Scale"*—or more accurately, the *"Beta Awareness Scale"*—emphasizes specifically the *beta* range of the *Standard Model*, the degrees[25] between (0.0) and (4.0); as does our simplified version of it, called the *"Emotimeter."*

THE "EMOTIMETER" & "AWARENESS SCALE"

The *(Beta) Awareness Scale* demonstrates all degrees of personal ZU that fall between two states: a fully expressed total *"Self-Actualized"* beta-Awareness at (4.0); and organic death of the physical body at (0.0). It as at (4.0) that we refer to a *Seeker* as having reached a *defragmented*[26] state equivalent to

---

24 **organic** : as related to a physically living organism or carbon-based life form.

* Grade III—"Mardukite Systemology"

25 **degree** : a physical or conceptual *unit* (or point) defining the variation present relative to a *scale* above and below it; any stage or extent to which something *is* in relation to other possible positions within a *set* of *"parameters"*; a point within a specific range or spectrum; in *NexGen Systemology*, a *Seeker's* potential energy variations or fluctuations in thought, emotional reaction and physical perception are all treated as *"degrees."*

26 **defragmentation** : the *reparation* of wholeness; a process of removing *"fragmentation"* in data or knowledge to provide a clear understanding; applying techniques and processes that promote a

the most basic level of personal *Actualization* in *Self-Honesty*, which is *"beta."* This is actually what we are "testing" with a *BAT*—and evaluating with *SET* worksheets.

The easiest way to understand this approach—and integrate an understanding of the *Awareness Scale* as applied knowledge —is by first introducing the subject of the *"Emotimeter."* What we are calling the *"Emotimeter"* is essentially a simplified version of our complete *(Beta) Awareness Scale*.

The original *Awareness Scale* marks 35 (up to 42) different frequencies or *degrees* on the ZU-line concerning THOUGHT and EMOTION. In contrast, the *Emotimeter* distinguishes only 9 (up to 21) different points. Both methods: provide the exact same function; describe the exact same conditions or states; and plot *degrees* of ZU simultaneously on the same ZU-line. The only difference is the refinement or clarity of data we might evaluate. One method is obviously quicker, more generalized and easier to commit to memory for immediate use; the other is obviously more specific, precise and useful for determining and predicting tendencies and patterns in detail.

UNDERSTANDING THE "EMOTIMETER"

Both the *Emotimeter* and *(Beta) Awareness Scale* may be applied to existing knowledge of the *"Standard Model of Systemology"* and the *"ZU-line"* relayed in a previous volume.[‡]

The *Emotimeter* can be used to evaluate a considerable wealth of information even without incorporating further divisions and descriptions charted by the full *Awareness Scale*. The previous *Assessment Scale* given to evaluate *BAT* questionnaire

---
holistic interconnected *alpha* state, favoring observational *Awareness* of continuity in all spiritual and physical systems; in *NexGen Systemology*, a *"Seeker"* achieving an actualized state of basic *"Self-Honest Awareness"* is said to be *defragmented*.

‡ *"Tablets of Destiny: Using Ancient Wisdom to Unlock Human Potential."*

statements is directly based on the *Emotimeter*. Therefore, a *Seeker* may use this to analyze basic data provided on a completed *SET* worksheet and addenda.

> NEXGEN SYSTEMOLOGY "EMOTIMETER"
> __4__ = Full Awareness Expressed ("Vibrant")
> __3.5__ = Outgoing/Pointed ("Positive")
> __3__ = Content/Friendly ("Casual")
> __2.5__ = Tolerant ("Dismissive")
> __2__ = Suspicious ("Pessimistic")
> __1.5__ = Violent/Spiteful ("Negative")
> __1__ = Evasive ("Afraid")
> __0.5__ = Grieving ("Sad")
> __0.1__ = Apathetic ("Unconscious")
> __0__ = Organic Death/Physical Universe (KI)

Some particular details became clear to us with our new understanding of *beta-Awareness*. For one: it became evident that the basic states described by the *Emotimeter* mirrored conditions of *"havingness"* and the attainment of *"knowingness"* reflected in traditional approaches to *"Self-Actualization"*[27] maintained by a few humanist-motivational[28] schools of contemporary psychology. It will also be recognized, by anyone with previous familiarity on such topics, that *NexGen Systemology* is advancing an understanding beyond what has already been laid out—with an "application" of knowledge[29] into a

---

27 **Self-actualization** : bringing the full potential of the Human spirit into Reality; expressing full capabilities and creativeness of the *Alpha-Spirit*.

28 **humanistic psychology** : a field of academic psychology approaching a holistic emphasis on *Self-Actualization* as an individual's most basic motivation; early key figures from the 20th century include: Carl Rogers, Abraham Maslow, L. Ron Hubbard, William Walker Atkinson, Deepak Chopra and Timothy Leary (to name a few).

29 **knowledge** : clear personal processing of informed understanding; information (data) that is actualized as effectively workable understanding; a demonstrable understanding on which we may 'set' our *Awareness*—or literally a *"know-ledge."*

more workable methodology[30] carrying effective idealistic goals for the "common man."

At the frequency degree of "personal fear" (1.0), an individual is only seeking information, or data from their environment, in order to simply *cope* with Reality in a primitive sense. At this level of *Awareness*, the motivation to "exist" is driven by resolving the most basic survival needs only. Of course, at this level of *Awareness* an individual is less likely to actually be successful in having these needs met by their own efforts and actions. In fact, this condition can be extended all the way up to (2.0). This is because the ability to *be, know* and *act* as *Self-directed* from the "I" as the *Alpha "Self"* significantly diminishes the "further down" in degree we are on any *Scale* based on the *Standard Model*.

Assuming basic needs are met, a person at (2.0) is no longer maintaining an active state of fear or anger toward their environment. They may therefore look to fulfill the next level of basic need: *security*. In essence, once an individual figures out how to establish its basic needs, the next logical effort is to maintain them—and so they seek the information and resources from their environment that will *assist* them in achieving a foundation of stability. Such stability promotes further expansion and longevity of personal existence.[31] At its most fundamental and biological core, this would include specifically *family* and our closest ties to other *Lifeforms* that *assist* us in our continued existence.

When the *Standard Model* was properly introduced,[‡] a suggest-

---

30 **methodology** : a system of methods, principles and rules to compose a systematic paradigm of philosophy or science.
31 **existence** : the *state* or fact of *apparent manifestation*; the resulting combination of the Principles of Manifestation: consciousness, motion and substance; continued *survival*; that which independently exists; the *'Prime Directive'* and sole purpose of all manifestation or Reality; the highest common intended motivation driving any *"Thing"* or *Life*.
‡ *"Tablets of Destiny: Using Ancient Wisdom to Unlock Human Potential."*

ion was presented that the average or "norm"—if we were to plot modern society as a whole in 2019—would be approximately between (2.0) and (2.5) on our *Awareness Chart*. This estimation still stands. It may even be evident in a *Seeker's* personal *BATs* prior to extensive *Systemology Processing* and education. Certainly an individual can rise above the "norm" for its society—and certainly the average for a society can change based on its individuals with an ability to radiate truth among the masses. We have also seen the effect of the other direction—such as a society at (1.0) or (1.5) in WWII Germany.

Application of *NexGen Systemology* at a *global* or *societal* level is outside the immediate scope of the current Grade of material —although additional volumes will undoubtedly support this information. For our current purposes we are concerned with treating and understanding the *individual*—yet in doing so we undoubtedly must present examples of which tend to turn our attentions to externals, particularly the world around us. A society between (2.0) and (2.5) is only interested in promoting and maintaining the most basic needs of an individual— which keeps the average population in a state somewhere between "pessimistic suspicion" and "dismissive tolerance" at best. Not very *enlightening*. The theme of society is: As long as people are not simply running amok, everything's fine.

We therefore do not see a drive toward any greater *enlightenment* or any larger effective group dynamic taking place until we reach approximately (3.0). It is at (3.0) that we begin to see signs of a healthy interest in *Life* and general existence. After physical conditions are relatively stable and there is a firm foundation for expansion, an individual's *beta-Awareness* is usually free and unwound enough from past and present concerns to pursue higher interests. This does not mean they will automatically discern only correct knowledge, or that they will be successful in all their endeavors, but the *interest* is there at a level we would consider at least somewhat *above* normal average.

An individual coming up to (3.0) for the first time is just beginning to truly *reason* for themselves. This individual could be book-smart, but they might also be overly cautious or conservative in accepting *any* information as facts. They might find many things interesting, but may still treat all data equally and casually. At the very least, a person at (3.0) is simply in the *know* that there is *something* "true" to *know*. They are becoming *Aware* that they are *Aware* and that there are things to be *Aware* of *Self-Honestly*. This seems trite—and yet it is considered above average to current societal norms.

If a *Seeker* is first discovering the *Mardukite* work, *Systemology* and the *Pathway to Self-Honesty* from subjectively lower points on the *Emotimeter*, it is when they elevate to (3.0) that they usually begin to notice the greatest *shift* between a point they were coming from and where they are headed. This state is sometimes achieved relatively quickly after applying basics of *Systemology Processing* and education for even a short time. The gradient "slope" between (3.0) and (4.0) is generally determined by the individual, their application of the principles and the amount of personal *fragmentation*[32] remaining to be resolved in the form of erroneous[33] "*emotional imprints*"[34] and

---

32 **fragmentation** : breaking into parts and scattering the pieces; the *fractioning* of wholeness or the *fracture* of a holistic interconnected *alpha* state, favoring observational *Awareness* of perceived connectivity between parts; *discontinuity*; separation of a totality into parts; in *NexGen Systemology*, a person outside a state of *Self-Honesty* is said to be *fragmented*.
33 **erroneous** : inaccurate; incorrect; containing error.
34 **imprint** : to strongly impress, stamp, mark (or outline) onto a softer 'impressible' substance; to mark with pressure onto a surface; in *NexGen Systemology*, the term is used to indicate permanent Reality impressions marked by frequencies, energies or interactions experienced during periods of emotional distress, pain, unconsciousness or antagonism to physical survival, all of which are are stored with other reactive response-mechanisms at lower-levels of *Awareness* as opposed to the active memory database and proactive processing center of the Mind; an experiential "memory-set" that may later resurface—be triggered or stimulated artificially—as Reality, of which similar responses will be engaged automatically.

"thought-formed[35] beliefs."

At higher levels of *beta-Awareness*—such as from (3.5) to (4.0)—an individual has already managed their past *imprints* and is managing present issues quite well. There have been many successful individuals and personalities in history that appear to regularly maintain these levels innately. This is what allows true *visionaries* and *leaders* to set their attentions and *Awareness* on the *future*, rather than being fixed to past or present problems. As such, we tend to find a motivation for *empowerment* occurring at this level—whether or not it is supplemented by *Self-Honesty*.

ACTUALIZATION OF BASIC NEEDS

Systemology education and higher-level *processing* techniques become increasingly critical for development when a *Seeker* experiences higher rates of personal vibration[36] or *ZU-frequency*. Once a *Seeker* has rehabilitated their identification (*realization*) of the *Alpha Spirit* or *Self*, the goal is quite simply to make the able *more able*. Up to this point in the program—as demonstrated within the present volume—the method is simply for a *Seeker* to work a little further and a little further to essentially achieve this baseline *Awareness* where "the real work"—the actualized *Alpha* work—can begin. It is at a point of "free expression" that we can focus on *what* exactly we are expressing. Because this is important—hence: education.

A person could be elevated to a point of sheer outward vibrancy and charisma *without* actually achieving a true *beta*

---

35 **thought-form** : apparent *manifestation* or existential *realization* of *Thought-waves* as "solids" even when only apparent in Reality-agreements of the Observer; the treatment of *Thought-waves* as permanent *imprints* obscuring *Self-Honest Clarity* of *Awareness* when reinforced by emotional experience as actualized "thought-formed solids" ("*beliefs*") in the Mind.
36 **vibration** : effects of motion or wave-frequency as applied to any system.

state of *Self-Honesty*.* The two states are clearly not the same; they simply resonate similarly by external observation.

If we consider "basic needs" data from Abraham Maslow's famous *"Pyramid of Self-Actualization"* in relation to (1.0) to (4.0) on the *Standard Model*—

> DYNAMICS OF BASIC NEEDS & BETA SURVIVAL
> __4__ = Humanity; success; personal esteem.
> __3__ = Group interaction; purpose.
> __2__ = Family unit; domestic security.
> __1__ = Self; physiology; primitive survival.

—we can arrive at many effective parallels between. An individual generally occupies a *level* of *beta-Awareness* consistent with a level of the Physical Universe (KI) they have achieved personal *mastery* of. By this *"mastery"* we *do not* mean some exercise of blatant (and ignorant) totalitarian control. We mean a level of *"mastery"* in the same fashion that an artist *masters* their craft. And we can be certain that the "crafting of universes" *is* an art-form (as much as it is cold science and logic) of which we are participating in by our agreements[37] of Reality at every moment. This is the knowledge that constitutes our *Standard Model*—deciphered from secrets kept in the oldest libraries of Mesopotamia, forged at the inception[38] of modern civilization and the original systematization of the modern *Human Condition*.[39]

In fact the *Standard Model* and our "quantification" (or *numeric values*) of the *ZU-line* would not be nearly as useful to us if it could not be directly related to other current valid data somewhere within its range to "know" things. This is because our *Standard Model* is a *"continuity model,"* which has been

---

* In *Systemology*, this is called "Grandeur" or the *"false Four."*
37 **agreement** : unanimity of opinion; an accepted arrangement; "reality."
38 **inception** : the beginning, start, origin or outset.
39 **Human Condition** : a standard default state of Human experience.

demonstrated in a former volume‡ to include *Everything* and *Nothing* from *Infinity-to-Infinity*. It is evident by our application of the *Emotimeter* to states of the Human Condition, that this model is applicable in demonstrating both the *subjective* and *objective* Universe—especially as we experience it in *beta existence*—and that it reflects a perfected map by which a person might navigate the *Game* of this Physical Universe and reach the *Pathway of Self-Honesty*.

DEVELOPING SCALES TO CHART SYSTEMS

The *Emotimeter* simplifies the more complete *(Beta) Awareness Scale*, developed from an intensive understanding of the *ZU-line*—when metered or gauged alongside the *Standard Model*. Our efforts did not result from *complicating* knowledge into a larger more complete scale; we developed the deepest understanding possible and then reduced it to a simplification. Our "true understanding" as "true knowledge" should always be in the direction of reduction: a wide range of valid data brought to a point or spark of *true knowing*. There is no reason to over-complicate a system unless its designers are set out for *authoritarian control*—in which the artificial convolution is allegedly only understood by a select few *"authorities."*

Our own intentions within *Mardukite Systemology* are only to distribute a complete holistic[40] understanding as it has been, and continues to be, discovered; *Graded* or *tiered*[41] by educational levels only so it may be realized into "true knowledge" with the greatest certainty, and not simply in some esoteric[42] attempts at cryptic delivery. The *"Ancient Mystery School"*—from which our founding data is drawn—also conducted itself

---

‡ *"Tablets of Destiny: Using Ancient Wisdom to Unlock Human Potential."*
40 **holistic** : the examination of interconnected systems as encompassing something greater than the *sum* of their "parts."
41 **tier** : a series of rows or levels, one stacked immediately before or atop another.
42 **esoteric** : hidden; secret; knowledge understood by a select few.

in a cumulative manner, carrying *Seekers* through initiatory degrees. In essence, the best route to learning is often conducted in this way—so long as the material is applicably *graded* specific to the individual. This is not always easy to arrange as a "static book." Grade-I and Grade-II apprenticeship programs (under an authorized *Mardukite Master*) are also currently in development—in addition to a *Piloting Program*[43] aimed at preparing professionals to deliver "Grade III"[*] (and above) *NexGen Systemology Processing*. All of this is aimed toward greater quality assistance to *the individual* as this movement and organization progresses into the future with its goals of global *spiritual transhumanism* leading toward a new Human evolution: the state of *Homo Novus*.

In addition to the *Standard Model*, both the *Emotimeter* and *(Beta) Awareness Scale*[‡] are useful tools and utilities—part of the *map, compass and square* assortment of every *Seeker*, *Systemologist* and *Mardukite Zuist*[*] on the *Pathway to Self-Honesty*. These tools not only assist in providing some of the deepest insights about ourselves—our *Self*—which is the foremost purpose of this methodology, but it also directly leads to a greater understanding of *others*, and most effectively, our interpersonal relationships and communication with others. This is important not only in facilitating the highest success and satisfaction in our *beta existence*, but also in nearly every aspect of *world-building* and *reality engineering*[44] managed at

---

43 **pilot** : the steersman of a ship; in *NexGen Systemology*—an individual qualified to operate *Systemology Processing* for other *Seekers* on the *Pathway to Self-Honesty*.
* "*Grade III—Mardukite Systemology*" is demonstrated in the present book.
‡ Which is itself only a portion of the "*Systemology Total Awareness Chart*."
* "*Mardukite Zuism*"—a Mesopotamian-themed (Babylonian-oriented) religious philosophy applying the spiritual technology based on *Arcane Tablets* in combination the "Tech" from *NexGen Systemology*.
44 **engineering** : the *Self-directed* actions and efforts to utilize knowledge (observed causality/science), maths (calculations/quantification) and logic (axioms/formulas) to understand, design or manifest a solid structure, machine, mechanism, engine or system; as "*Reality*

the highest levels of personal *Awareness* as the "I" or *Alpha Spirit*.

For those *Seekers* continuing on to this material from *Liber-One*,[∞] it will be noted that "*Processing*" in this book is primarily concerned with actualizing the portion of the *ZU-line* (and *Standard Model*) that directly relates to *beta-existence* in the Physical Universe—sometimes referred to as "KI" in *Mardukite Zuism*, as derived from a Sumerian *cuneiform* sign for the "material world" and planet Earth. This *beta* range of personal *ZU-frequency* vibration interacting with the Physical Universe (KI) as *Life in Awareness (ZU)*, is plotted between two points:

a.) the point of densest most inert physical matter at (0.0) or *KI*; and

b.) the point at which the *Alpha Spirit* of the "*I*" of-and-as "*Self*" contacts systems of the *genetic vehicle* that it controls, or is anchored to, at (4.0).

THE SYSTEMOLOGY OF SPIRITUAL IMMORTALITY

Between (0.0) and (4.0) are various *beta states* of *fragmentation* inherent in a *Seeker's* participation in *beta-existence* as the Human Condition. These *genetic vehicles* are actually quite amazing bodies to have use of while interacting with the Physical Universe (KI). They may not be necessary, but they are convenient to have use of while we work up to the research and actualized states that will allow us to move between them *at will*. Unfortunately, without this knowledge, as an individual becomes more and more the *effect* of this

---

Engineering" in *NexGen Systemology*—intentional *Self-directed* adjustment of existing Reality conditions; the application of total *Self-determinism* in *Self-Honesty* to change apparent Reality using fundamentals of *Systemology* and *Cosmic Law*.

∞ "*Tablets of Destiny: Using Ancient Wisdom to Unlock Human Potential.*"

Physical Universe—and thereby less of a *Self-Honest Cause* to Reality—all *fragmentation* thickens, erroneous *Imprints* become deeper and more numerous and *thought-forms* become more solid.

This phenomenon tends to happen with "age"—as an individual agreeing to the Human Condition accumulates more and more of this *reactive programming*. This can also happen throughout the course of a lifetime with the accumulation of painful experiences—and potentially even carried through many lifetimes. As a result, *Awareness* moves down the *Emotimeter* or *Scale*—primitive "reactive behavior" concerning basic needs and securities of preserving physical existence begin to take over and displace[45] *Alpha Awareness*. An individual first begins to *react* "as a wild, sick or injured animal"—and then at even lower levels of *Awareness*, falls into an apathetic "near death" state of hopelessness.

The personal states described on the *Emotimeter, (Beta) Awareness Scale*—or directly on the *ZU-line*—concern *objective* frequencies and vibrations that are experienced *subjectively*. They are derived from basic logic[46] and observed universal mathematics, not *"existentials."* This means that the highest band[47] of *beta-Awareness* between (3.8) and (4.0) could include any states reached as personal "peak experiences" of a *Seeker's* lifetime—points of greatest achievement and milestones of utmost success that are the apex of human experience and "ecstasy" when we are "walking on air" seemingly "outside of ourselves" (*ex-stasis*) &tc.

---

45 **displace** : to compel to leave; to move or replace something with something else in its place or space.
46 **logic (equations)** : using symbols and basic mathematical logic to establish the validity of statements or to see how a variable within a system will change the result; a basic demonstration of proportion or relationship between variables in a system.
47 **band** : a division or group; in *NexGen Systemology*—a division or set of frequencies on the ZU-line that are tuned closely together and referred to as a group.

We say that our quantitative (*numeric approach*) to *beta-Awareness* does not concern "*existentials*" because it says nothing directly of whether or not *Self-Honesty* is *actually* maintained at these peak points—or of an individual's ability to maintain this "peak elation" as *fully Actualized beta-Awareness* thereafter. Of course we are not seeking some "external event" to spark a change in us—but the "peak experience" similarity is the best relative equation we may provide to a *Seeker* at this time. In essence, a *Seeker* that is elevated to higher levels of *beta-Awareness* would have resolved their personal management of the basic securities related to material "*Havingness*" and would therefore be moving their attentions[48] to higher demonstrations of Reality activity *realized* in the band of thought vibration as *Knowing* or a state of "*Knowingness*"—which is not the same as the sheer accumulation of knowledge, but rather a *reduction* of data, facts, imprints and experiences into *effective truth*.

Although the *Emotimeter* and *Awareness Scale* pertain specifically to personal states of *beta-Awareness*, the *Seeker* will not be "kept in the dark" concerning higher level potentials inherent within our *Systemology*. When the logic of personal *ZU-frequency* is applied to higher *Alpha* levels, new theoretical avenues open up to us on the *Pathway to Self-Honesty*—even beyond the achievement of an *Actualized* defragmentation of *beta-Awareness* as *Self* at (4.0), including:

—true understanding and mastery of material
 Life-Existence (5.0);

—true understanding and mastery of *Universes, Games & Logics* of spirit and matter (6.0);

—total *Self-Actualization of "Spiritual Life-Beingness"* (7.0); and

—actualized *Self-transcendent "Infinite Beingness"* at
 Source (8.0).

---

48 **attention** : use of active *Awareness* toward a specific aspect or thing.

SCALES AND DEGREES ON
A CONTINUITY MODEL

An individual's *Lifeforce (or ZU)*, experienced as *Spiritual Awareness* of *Self*, is a constant force of *Will* set in action from a Source in Infinity. A lessening of total *Awareness*—and equally the actualization of *Self-Honesty*—in the experience of this *Lifetime*, is only a result of two things:

    a.) turbulent emotional energy wound up in *encoded imprints*; and

    b.) creative mental energy cemented in *thought-formed beliefs*.

The first (a) is primarily resolved with intensive *Systemology Processing* combined with *Self-Evaluation*; the second (b) is handled the same, combined with extensive *Systemology Education*, so that the reality of *thought-formed beliefs* and *encoded imprints* is better understood and managed in the future. The end goal of *Mardukite Systemology* is always in the direction of "*Self-reliance*" and not to promote some misguided emotional dependency on any religious organization or spiritual institution.[49] Such methods were once tried by all carriers of some truth in the past; and as a result, all failed to deliver what they promised their *Seekers*.

One of the primary axioms of *NexGen Systemology* and *Mardukite Zuism* is that:

"**The *Self* occupying the *genetic vehicle* is a 'spiritual universe cause' of 'physical universe effects' engaging a *Self-determined* Alpha-Spirit 'Will' as *Actualized Awareness* in beta existence.**"[*]

Consequently we can rate or quantify the *degree* or *level* of

---

49 **institution** : a social standard or organizational group responsible for promoting some system or aspect in society.

* Excerpting fundamentals from "*Mardukite Zuism: A Brief Introduction*" and "*Tablets of Destiny*" by Joshua Free.

BETA-AWARENESS ACTUALIZATION %
4 = 100%
3.5 = 70%
3 = 50%
2.5 = 35%
2 = 25%
1.5 = 15%
1 = 10%
0.5 = 5%
0.1 = 1.5%
0 = 0.0%

personal actualization chronically expressed in *beta-existence*—on the *(Beta) Awareness Scale*—if we apply "percentage values" from the *(Set) Analysis Graph*.‡

Awareness is the best word we can used to apply to *ZU*. It is the definitive *Self* that is "conscious" as "I" when controlling a physical body; the *Self* that is *Willing* all sense of *doing, having, knowing* and *being*; and ultimately the *Alpha-Spiritual Self* that remains "I" even after a genetic death of the physical body it maintains control anchors with. *Awareness* is then more closely tied to our concept of personal *existence* than any other available. Hence the exhaustion of the use of the term throughout this book—and also *Self-Honesty*, which is the actualization of that *Awareness* free from inhibitory and reactive programming that displaces *Self* in managing interpretations and agreements of Reality.

The important thing to keep in mind whenever examining *degrees* and the *levels*[50] of *Awareness* described within specific

---

‡ This is estimated with mathematical logic within the *Beta* range from the *Standard Model*. The full continuum for *"Infinity-to-Infinity"*—which must include everything *exterior* to *beta existence*—would extend not only from (8.0) to (0.0), but then also down to (–8.0). Personal experience of the "Physical Universe" only manifests within vibrations of the levels between (0.0) and (4.0), but this does not even exclude the possibility of other "Physical Universes" also existing at that frequency, which are spatially separated. The *Standard Model* does not actually prove this, since it is established from the perspective of "Self" in *this* "Physical Universe"—but it does not exclude the possibility. More of these considerations, which at this time may seem distracting, will be reserved for higher *Grades* of material.

50 **levels** : a physical or conceptual *tier* (or plane) relative to a *scale* above and below it; a significant *gradient* observable as a *foundation* (or surface) built upon and subsequent to other levels of a totality or whole; a *set* of *"parameters"* with respect to other such *sets* along a

parameters,[51] is that these are *scales* within said *parameters* and not systems being treated in isolation or exclusion to all other *macro-systems* and *micro-systems* active "above" and "below" wherever we are examining on the *ZU-continuum*.[52] For our current purposes, we have chosen to focus our attentions specifically on factors directly related to *beta-Awareness* and our basic management of *Beta Existence*—the "Reality" of the Physical Universe (KI).

Another important feature of the *Awareness Scale* is that it is a *subjective* representation of *objective* energy vibrations. It relates very precisely to general interactions between frequencies, but not the actual content communicated or complete validity of data transmitted. Yet we can be certain, for example, in the instance of an *Alpha-Spirit's* ability to form *imprints* that:

a.) lower levels relate to states of emotional turbulence and *emotional encoding* that generate those states of a destructive nature, contributing to a rejection of the Physical Universe; and

b.) higher (*beta*) states promote a personal motivation toward material success and achievement accomplished through effectively maintaining and transforming the Physical Universe with *thought-forms*.

We can apply many such "dichotomies" to the range of any scale we choose to apply to our *Standard Model*, so long as we

---

continuum; in *NexGen Systemology*, a *Seeker's* understanding, *Awareness* as *Self* and the formal grades of material/instruction are all treated as "*levels*."

51 **parameter** : a defined range of possible variables within a model, spectrum or continuum.

52 **ZU-continuum** or **ZU-line** : a spectrum of *Spiritual Life Energy*; an energetic channel of *Identity-continuum* connecting the *Awareness (ZU)* of an *Alpha-Spirit* with "*Infinity*"; a *Life-line* on which *Awareness (ZU)* extends from the direction of the "Spiritual Universe" (AN) as its *alpha state* through an entire possible range of activity in its *beta state*, experienced as a *genetic-entity* occupying the *Physical Universe (KI)*.

understand that it is not a *finite dualistic* model, and that it actually extends in *degree* toward *Infinity* in directions or levels both "above" and "below" wherever we fix the parameters of a scale. For example, if we were to theoretically scale *Self-Determinism* of an *Alpha Spirit* on the *Standard Model*, it would run from 100% Infinite Cause (8.0) down to 100% Physical Effect (0.0). Of course, we only experience a portion of this spectrum[53] within the *beta* range as *Awareness* using existing faculties of the Human Condition. Therefore, midway between we find the interaction between the *Alpha* and *beta* at (4.0), which is the key point of *Self-Honest* transmission between the *Alpha Spirit* and the Mind-Systems interacting with a *genetic body*.[∞] We could apply similar scales of "Right and Wrong" and "Always and Never"—or of any *positive* condition (its state of *"beingness"*) or creation and at the other end, its state of *not-being*. This becomes increasingly evident with examination and study of the *Awareness Scale*.

USING THE (BETA)-AWARENESS SCALE

As a *Seeker* increases their familiarity with various fundamentals of the Human Condition, they discover more about themselves and others—but as a variety of knowledge that leads to *"true understanding,"* which must (by definition) also lend itself toward a mostly valid prediction of future results based on existing evaluations. Without such consistency, all we are left with is a vague idea—nothing of substance to base our *model* upon. Fortunately we have developed a strong *model* for our purposes, drawn out from thousands of years of humanity's research concerning *beta-experience* in addition to our own applications and experimental pursuits of *NexGen Systemology*.

---

53 **spectrum** : a broad range or array as a continuous series or sequence.
∞ This point at (4.0) is quite properly referred to as the *Master Control Center* or *MCC* in *Liber-One*, *"Tablets of Destiny"* by Joshua Free.

Understanding and applying the *(Beta) Awareness Scale** allows an individual the best chance of understanding themselves and individuals around them. In addition we earn a greater understanding concerning our environment and our ability to manage it successfully. Some of these details become clearer—or with greater "reality"—during personal *processing*.

The fundamental *spectrum* reflected in the *Emotimeter* and *Awareness Scale* is a description of the most basic *levels* of *beta-existence*—and personal *Awareness* of the same. The ideal basic state of the Human Condition operating from the *Alpha-Spirit* is marked at (4.0) on these charts. This is the point on the *Standard Model* that differentiates a frequency threshold[54] between two distinct states:

a.) energy and matter of a *"Beta"* Physical Universe (KI); and

b.) energy and matter of an *"Alpha"* Spiritual Universe (AN).

On a personal level of existence, when we are dealing with degrees in the range of *beta*, we are referring not so much to the frequencies of our environment, but those that relate specifically to our own expression and interaction (of ZU) with environmental energy. Thus we can better understand, for example, an individual communicating from any particular level of vibration that can produce effects in the environment and on others in that environment.

We see a range of physical interaction between (0.0) and (1.0) that is reduced to the most dense inert low energy physical matter. We can chart other emotional interactions between (1.0) and (2.0) that remain below the level of thought and reason and are primarily reactive-response mechanisms developed from *emotionally encoded imprints* of experience. Between (2.0) and (4.0) we plot all degrees of *"Thought"* activi-

---

* Hereafter referred to simply as the *"Awareness Scale"* in this text.
54 **threshold** : a doorway, gate or entrance point; the degree to which something is to produce an effect within a certain state or condition.

BETA-AWARENESS SCALE CHART

"4.0"	SELF-HONESTY (BASIC STATE)	(4.0) Charismatic
"3.9"	SUCCESS/ACHIEVEMENT	<
"3.8"	Elated/"In Love"	(3.8) Enthusiastic
		(3.6) Cheerful/Energetic
"3.5"	CONFIDENT	<
		(3.4) Determined/Eager
"3.2"	Strong Interest	(3.2) Vigorous/Alert
"3.0"	INTERESTED	<
"2.8"	Mild Interest	(2.8) Encouraged
		(2.6) Doubtful/Disinterest
"2.5"	INDIFFERENT	<
		(2.4) Bored/Tired
"2.1"	Monotony	(2.2) Neglect/Dislike
"2.0"	INVALIDATING	<
"1.8"	Pain Sensation	(1.8) Antagonism
		(1.6) Confrontation/Violence
"1.5"	ANGER	<
		(1.4) Hateful
"1.2"	Resentment	(1.2) Anxiety
"1.0"	FEAR	<
"0.8"	Numb	(0.8) Terror
		(0.6) Grief/Loss
"0.5"	SUFFERING	<
		(0.4) Depression
"0.2"	Victimization	(0.2) Hopelessness
"0.1"	APATHY	<
	Pretend Death	(0.05) Uselessness
"0.0"	INERT PHYSICAL UNIVERSE	(0.0) Organic Death

ty and the expression of *Self* in the Physical Universe (KI). It is within *this* range of "Mind-Systems" that most *Processing* within this present volume is targeting.

This point of *Self-Honesty* we are *Seeking* in our work is not some newly concocted state that we have decided upon. It is, quite the contrary, our *truest* state—and quite attainable in a lifetime. It is actually imperative that one does—or is at least consistently developing in that direction on the *Pathway to Self-Honesty*, because it is very evident when we examine our *models* and the *Awareness Scale* that a "higher state" correlates[55] directly with a "higher ability" to not only manage *Self* in *beta-existence*, but ensure the most optimum continuation of the *existence of Self*. In the most primitive terms: our *level* of *Self-Honesty* is directly linked to our *level* of survival and thus our ability to manage basic needs as a foundation to succeed at greater personal achievements.

The *Awareness Scale* is concisely drawn out from research and experimental data that would fill many superfluous[56] volumes in itself. We have simply selected the most appropriate terms that best reflect each *degree*—although many other "keywords" and "associations" undoubtedly may apply. A *Seeker* that has understood the *Emotimeter* will already be familiar with the basic systematic flow demonstrated on the *Awareness Scale*—however, the complete scale divulges the "blue-prints" or inner workings of the "*ZU-fluctuation*" that result in various states, and naturally for our purposes, how they are systematically linked together. For example, you will discover with each basic state that further *adding* or *subtracting* even a little bit of *Awareness* will promote movement toward another related state. A *Seeker* should spend some time studying the *Awareness Scale* before continuing.

---

55 **correlating** : the relationship between two or more aspects, parts or systems.
56 **superfluous** : excessive; unnecessary; needless.

UNIT TWO
SELF PROCESSING

"SELF-PROCESSING" IN SYSTEMOLOGY
(DEFINED)

Self-Processing is an *application* of traditional *Systemology Processing* performed on one's *Self* alone. It is a specialized form of *self-directed attention* and *focused concentration* using materials within the field of *Systemology* as a guide. Performing this Systemology work alone *is real* processing, even when unaided by a friend or experienced *"Pilot"*[57] of these procedures. All *Self-Processing* indicated in this book is directed toward a specific systematic goal. This is far different from:

Personal "Self-Talk" (the voices and dialogue often running in our heads);

"Free-wheeling" (the random thought review and light recall experienced regularly when not fully directed); or

"Solo-Piloting" (experienced use of advanced *Systemology* applications to one's *Self* without additional aid).

One version of traditional *Systemology Processing* technology—called *"Cathartic Processing"** within the text of *"Tablets of Destiny: Using Ancient Wisdom to Unlock Human Potential"*—goes through intensive procedures best performed (and even learned) while working with a partner or experienced Nex-Gen *Systemology "Pilot."* Technically and precisely speaking, even "piloted" *Systemology Processing* must be *"Self-Processed"* if it is in any way effective. Whether a *Seeker* is practicing entirely *Self-directed* with a *Systemology* book alone or with another individual using *Systemology* book-learning, to achieve *any benefit*: <u>all</u> *Systemology* processes must be effectively "processed" by *Self*.

---

57 **pilot** : the steersman of a ship; in *NexGen Systemology*—an individual qualified to operate *Systemology Processing* for other *Seekers* on the *Pathway to Self-Honesty*.
* "Systemology Process RR-SP-1"

The *techniques* and *education* within *Systemology* are combined to compose the *"Technology"*—and it is offered to you—the *Seeker*—as an effective "map" and workable "tool." But keep in mind: *the map is not the journey*; and a tool is only useful to the extent that may be effectively applied. There is no substitution for YOU making the effort, actualizing the journey as *Self*—and creating the most successful existence possible—applying the truest understanding of the tools and their applications at each level of *Awareness*.

On each solid *"ledge"* you have the potential to realize a certain state of *"Knowingness"* directly—and on this *certainty* of *"know-ledge"* you can extend your reach a little further, each time achieving higher and higher foundations of thought and existence. We must lay our own course, direct our own fate and build our own personal *Ladder of Lights* to ascend to our highest actualization of *Self*.

THE PURPOSE OF "PROCESSING" TOWARD SELF-HONESTY

The ultimate goal of all basic *Systemology Processing* is to effectively increase *Awareness* level of an individual *Seeker* on the *"Pathway to Self-Honesty."* The purpose of these basic techniques is personal *"defragmentation"*—or else clearing out *fragmented* distortions from our energetic expression of *Alpha thought* as the "cause" demonstrated in the Reality of *beta-existence*.

A *Seeker* maps their progress using *"Self-Evaluation"* as they increase *beta-Awareness* concerning *Self*, other lifeforms, the environment and ultimately, universes. As a result, actualized personal faculties and abilities to process and manage *Life-experience* increases. A graded level of personal vibration/frequency—such as reflected on the *ZU-line*—is related directly to any degree of *certainty* within reach as "I" or *Self*.

"Self-Processing" in the present volume emphasizes *analytical evaluation* of *experience* retained as memory. The Human Condition occupies a society that prides and gauges its own level of "certainty" strictly on the accumulation of experience in the form of *emotional imprints* and *thought-formed beliefs*. Whenever these *imprints* and *beliefs* are not brought to a scrutiny in the highest range of our *analytical thought*, our energetic potential (and range of actualized *Awareness*) is lowered. The personal *ZU-energy* and *Awareness* that would otherwise be used to "fully" process the *Present* and apply toward a fully Self-directed *Future* in *Self-Honesty* is otherwise being wound up in the *Past*.

It is the purpose of basic *Systemology Processing* to untangle and discharge personal stores, links, energetic ties, "*imprints*" and "*fragmentations*" that are not promoting the Seeker's "rise" in consciousness. *Cathartic Processing*—described as "RR-SP-1" in a former volume‡—target deep emotional turbulence directly. This rather intensive method, based on steps instructed on a fragment of an ancient cuneiform tablet—in combination with other esoteric practices—demonstrates the core sequence of effective basic processing when assisted by the *Awareness* of another individual.

Methods used in the present volume will still target encoded[58] *imprints*—since such is always a necessity when working toward personal *Self-Honesty*. But without the assistance of a *coach* or "*Pilot*" to prompt and support a person through various directions and keep a *processes* running completely through to its effective end, more of the responsibility falls onto the individual *Seeker* themselves to remain *Self-determined* in their approach to, and gauging of, *Self-Processing*.

---

‡ "*Tablets of Destiny: Using Ancient Wisdom to Unlock Human Potential.*"

58 **emotional encoding** : the substance of *imprints*; associations of sensory experience with an *imprint*; perceptions of our environment that receive an *emotional charge*, which form or reinforce facets of an *imprint*; perceptions recorded and stored as an *imprint* within the "emotional range" of energetic manifestation.

This means developing one's own personal approach to the work. Obviously in order to deliver a "standard" in book-form, specific suggestions for *Self-Processing* are often generalized. An experienced *Systemologist* or *Pilot* can more precisely tailor processing to an individual's case directly—however, because a *Seeker* is expected to work through the present volume multiple times, greater and greater certainty can be expected to occur with each cycle. Personal ability to apply the work properly to the *Pathway* solely on one's own direction increases in relation to the *Seeker's* education and experiences with effective progressive use of *Systemology*.

OPERATING "SELF-PROCESSING" TOWARD SELF-HONESTY

A solitary *Seeker* may fully *Self-direct* an application of all *Systemology Processes*—those included in this present volume and elsewhere—so long as they persist with their education and practice, maintaining the persistence and personal fortitude necessary to continuously move forward and see themselves through on the *Pathway to Self-Honesty*, solely on their own *Self-determination*.[59] A proper "systematized" approach to this progressive development is even more critical with the absence of direct assistance by "someone who has already been where you want to be." It is the purpose of this present volume to provide that assistance by its author. Our handbook is valuable not only to those walking the *Pathway* entirely on their own *Self-direction*, but also those assisted by others—and for official *"Pilot"* education, those professionally trained to work directly with *Seekers* using this same *Tech*.

This present handbook is systematically designed to provide the most significant results possible while operating within very general guidelines. To be effective, a *Seeker* must work *Systemology Processes* through completely, meaning so long as

---

59 **Self-determination** : freedom to act, clear of external control or influence.

they are effective in yielding a change in state. Most individuals that do not find significant results also tend to end a *Process* before its completed; meaning while they are *bored* or *uncomfortable*. This is why "freewheel thought" is not as effective in yielding permanent results. Once a state of increased *Awareness* is attained—demonstrable by a greater enthusiasm or extroversion for one's life—then a *Seeker* may end the present session or move on to another *process*. After moving through all relevant work provided herein, the processes should be revisited again in sequence. Each time a *Seeker* works though a cycle of processes, they are easier to run through later and yet each time reaching more advanced levels of true understanding.

Expect to cycle though this book completely *at least three times*—and then this preset volume will serve thereafter as a powerful reference when needed. With that in mind, you should feel confidant to move through the material at a regular pace to achieve a general increase in *Awareness* and then go back and work through it over again as a new cycle of development. You will quickly discover, if you apply yourself to the *Processes* and continue through the book at a steady pace, then a cycle back to the "same processes" from the beginning will actually not be the "same experience" at all.

An individual session process must be run for as long as it is still producing a change in state and no longer or shorter. The *Seeker* may descend first through a whole host of associated low-level emotional states when an *imprint* or *thought-form* is resurfaced—even if the intent is only on "recall" of "positive" experiences, there are often many other memories and *imprints* and emotional stores of energy that are attached to them. When this emotional energy is discharged fully, the content of the *imprint* or memory is able to be brought clearly into an elevated *analytic* state where the memory may then be treated for its content and validity—and not solely for its sensations and reactive conditioning. The power of the reactive conditioning dissipates as the energy is raised to higher frequencies within the range of *Self-direction* and *Self-determination*—well above the range of the *"Reactive Control Center"* (RCC)[60] at position (2.0) on our models, charts and scales.

All *Systemology Processing* at this level of instruction is engaging the *analytical* Mind-Systems that govern transmission of thought frequencies between (2.1) and (4.0). A *Seeker* should not target *imprints* and *emotional encoding* if they, themselves, are hovering around or below (2.0) on the *Awareness Scale*. This is because the amount of personal ZU energy (*Awareness*) required to effectively resurface, confront[61] and reduce heavy programming and conditioning must exceed the energetic level of where the energy charge for the *encoding* is stored.

An individual maintaining a loosely angry or antagonistic state in the present—or maintaining some other anxiety about a present problem or issue—will be unable to sufficiently discharge *imprints* and energies held down near that frequency. Such a condition must be resolved as a part of the *"Self-Processing"* session *prior* to operating any *Tech* intended to reach higher levels *analytically*. When a *Seeker* is concerned about something other than the session, whether alone or with a *Pilot*, they are not "present" *in* the session—they will not receive the benefits of any *Processing*. This is, perhaps, one of the chief reasons for ineffective results—again, whether alone or with a *Pilot*.

---

60 **Reactive Control Center (RCC)** : the secondary (reactive) communication system of the *"Mind"*; a relay point of *Awareness* along the Identity's *ZU-line*, which is responsible for engaging basic motors, biochemical processes and any *programmed automated responses* of a living *beta* organism; the reactive Mind-Center of a living organism relaying communications of *Awareness* between causal experience of *Physical Systems* and the *"Master Control Center"*; it presumably stores all emotional encoded imprints as fragmentation of "chakra" frequencies of *ZU* (within the range of the *"psychological/emotive systems"* of a being), which it may *react* to as Reality at any time; in *NexGen Systemology*, this is plotted at (2.0) on the continuity model of the *ZU-line*.
61 **confront** : to stand in front of; to meet "face-to-face."

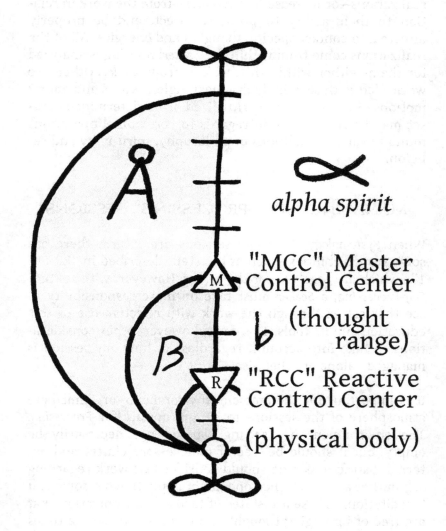

All *Systemology Processing* is arranged around a systematic "direction of attention." A *Seeker* only receives true *analytical* realizations—or increase in *Awareness*—from the work in relation to their ability to properly direct, and be properly directed, to contact specific thoughts and energies. All of the realizations come from within the *Seeker*; nothing is analyzed for them—either within this book or from a *Pilot*. Otherwise we are left with seemingly dry meaningless words and empty motions—just as any other ritualized or predetermined dataset may be reduced to, in regards to ceremonial operations found in various traditions of philosophy, spirituality and religion.

MANAGING "SELF-PROCESSING" SESSIONS

When *Systemology Processing* sessions are *Piloted*, there are some responsibilities—such as the steps described in "RR-SP-1"—which are resolved for the *Seeker*. However, with solitary *Self-Processing*, a *Seeker* must take on more responsibility to see themselves through the work with effective use of this education and its tools. There are, however, other considerations to take into account, regardless of how *any* session is managed—alone or otherwise.

It is important for the ambient environment—or immediate atmosphere of the session—to be appropriate for *Processing*. The environment or "setting" should not necessarily be empty, but it should be free of unnecessary clutter and external distractions. One should consider this work regarding *Self* in the same reverence one might apply to what some call "meditation." These are states of focused concentration that are free of impinging thought and emotion—these are times of *clear awareness*, and they are sought (for good reason) in every effective mystical, spiritual or religious practice. In fact, it is the personal result of these states—far more than on the shoulders of external technologies—that true human evolution takes place, and the ability to operate in *Self-Hones-*

ty increases. The condition of *Homo Novus*[62] will certainly include the increased capacity of actualized *Awareness*. The total extent of specific potential at these higher states may only be treated theoretically at this Grade* of materials.

There is another important consideration when *Processing*: the physical condition of the *Seeker* themselves. One common technique—before the session even begins—is to *stop*, HALT, and determine if you are (or the *Seeker* is):

Hungry, Aggravated, Lethargic or Thirsty.

An individual should never be processed under these conditions. There are many other recommendations concerning the general physical well-being of a *Seeker* and their health regimens in a future volume regarding ZU-energy,‡ but suffice to say that a person should, at the very least, be at (2.0) or above on the *Awareness Scale*, comfortable, clean, well rested and fed—essentially all "primitive needs" met—before any intensive *Processing* sessions. However, in the instance of *Self-Processing*, a light use of *analytical recall* methods may actually be used on "high-frequency" memories of personal success to—at least temporarily—raise personal *Awareness* enough *to be "in session"* fully and receive benefit from *Processing*.

---

62 **Homo Novus** : literally, "new man"; the "newly elevated man" or "known man" in ancient Rome; the man who "knows (only) through himself"; in NexGen Systemology—the next spiritual and intellectual evolution of *homo sapiens* (the "modern Human Condition"), which is signified by a demonstration of higher faculties of *Self-Actualization* and clear *Awareness*.

* Referred to as *"Grade III–Mardukite Systemology."*

‡ Released as *"The Power of Zu: Keys to Increasing Control of the Radiant Energy in Everyday Life"* by Joshua Free, based on "The Power of Zu" Systemology Conference lectures (December 2019).

AN INTRODUCTION TO KENOSTIC PROCESSING—"AR-SP-2" [EXTENDED COURSE]

This is a transcript of a lecture given by Joshua Free on the evening of October 18, 2019. Content of this lecture introduces course material retained by the NexGen Systemological Society (NSS) from the International School of Systemology (ISS). It is included within this volume to supplement previous chapter-lessons and for the benefit of Seekers (and Pilots) interested in experimenting with practical applications of Mardukite Zuism and Systemology (ZU)-Tech alluded to throughout this volume. Additional textbooks, guides and manuals are currently in development to advance further work within this applied spiritual philosophy of Systemology.

∆ ∆ ∆ ∆ ∆ ∆ ∆

Just over two months ago, I presented some history and a basic formula for *Cathartic Processing* as "RR-SP-1" at the *"Tablets of Destiny Convocation of Systemology."** This method, when *Piloted*, is quite effective in directly contacting and discharging emotional energy wound up in an individual. There is no doubt about that. However, this evening I am introducing material regarding a new "self-help book" that is currently in development‡—which is emphasizing the capabilities an individual has in using *Systemology* education and technology to achieve progress on the *Pathway to Self-Honesty* independent of any other direct assistance. Obviously some people are able to gain greater benefit than others without an experienced "Pilot" or even a friend to coach them. In any case, the former methods of *"RR-SP-1"* are not necessarily the most appropriate for solitary *Self-Processing*. Intensive catharsis relies on an individual successfully reliving dramatic

---

* *Cathartic Processing* lectures given on August 9, 2019; transcripts are included in the volume: *"Tablets of Destiny: Using Ancient Wisdom to Unlock Human Potential"* by Joshua Free.
‡ The present handbook you hold in your hands.

reenactments out properly to effectively discharge the emotional content of an *encoded imprint*. Otherwise, all a *Seeker* has done is successfully "restimulate" them—and we tend to call those folk "drama queens."

Methods introduced for *Self-Processing*—even if a *Seeker* is assisted in their use—are not intended to blatantly "restimulate" or "resurface"[63] the effects or affects of an *imprint*. But, since this can sort of just *happen* whenever one is working with *defragmentation*, the subject of "emotion" is emphasized in a former lecture series on *emotionally encoded imprints*. It is important information to know about. Our attention now is going to turn away from *Catharsis* to another similarly ambiguous ancient spiritual practice that the Greeks called *"Kenosis,"* thereby lending to our methods the name *"Kenostic Processing."* The information relayed this evening will be specifically pertaining to this subject of "Kenosis" and its effective application for our purposes as *"Systemology Process AR-SP-2."*

△ △ △ △ △ △ △

The word *"Keno"*—otherwise from the Greek *"Kenos"*—is tied to the old *Proto-Indo-European*[64] root: *"ken-"* meaning essentially "empty." This is applied to our *Processing* to correctly mean many things—and since each one of them is semantically correct, we would have no real need to codify[65] or systematize a whole series of "emptyings" or "emptiness" *processes*—but since this *is Systemology* we are talking about,

---

63 **resurface** : to return to, or bring up to, the "surface" what has been submerged; in *NexGen Systemology*—relating specifically to processes where a *Seeker* recalls blocked energy stored covertly as emotional *"imprints"* (by the RCC) so that it may be effectively defragmented from the *"ZU-line"* (by the MCC).
64 **Proto-Indo-European (*PIE*)** : a single source root language c.4500 B.C. contributing to most European languages.
65 **codification** : the process of arranging knowledge in a systematic form.

naturally we will end up doing that anyways, right?—in *Self-Honesty*. I mention this in this way because if you start to research what *Kenosis* means outside of *Systemology*, the waters get pretty murky.

The "spiritual emptying" as *Kenosis* has been applied in the past to two directions of travel for *Self* to experience itself in existence—or as an "effect" of its existence; or else *beta-existence*. In Christian mysticism, Jesus *empties out* his "divinity" in order to "become human." The true identity of his *Self* would remain unchanged in the Spiritual Universe—as does our own —but according to their definitions, this would be a form of *Kenosis*. We can approach it from the other direction as well— changing states toward the direction of our "Ascension"[66] right? If we just keep within the confines of this present example for a moment, we would at some juncture discover that this other *Kenosis* must take place for *Self* to now purge the Human Condition that it once chose to take on in order to return to the Spirit. However, without this high level of *Self-Actualization*, "body death" in and of itself is not any guarantee of actualized "Ascension"—and we are not advocating that as a Route. You'll shed that sucker when you are good and ready. It just so happens, however, when we look back in history that some individuals who do achieve these high states are *ready to go* and apparently as a result find themselves becoming iconic archetypes[67] of martyrdom and sacrifice.

Methods behind effective *Kenostic Processing* relate to essentially "analytical recall." The better a *Seeker* is able to handle "analytical recall," the more effective any other *Systemology Process* will be toward progressing them on the *Pathway to*

---

66 **Ascension** : actualized *Awareness* elevated to the point of true "spiritual existence" exterior to *beta existence*. An "Ascended Master" is one who has returned to an incarnation on Earth as an inherently *Enlightened One*, demonstrable in their actions—they have the ability to *Self-direct* the "Spirit" as *Self*, just as we are treating the "Mind" and "Body" at this current grade of instruction.
67 **archetype** : a "first form" or ideal conceptual model of some aspect.

Self-Honesty. We are still working within the domain of "subjective processes." Both *Cathartic* and *Kenostic* methods are entirely "subjective" from the point of the "Mind's Eye." There are other *Processes* that may be performed that are *objective*, eyes wide open, dealing with the space around us—but these tend to be more difficult to manage and direct properly as solitary *Self-Processing*. I do intend to include some in the "self-help book." But, for the moment, I have decided to focus specifically on "analytical recall" for *Self-Processing*. Hence: "AR-SP-2."

Routes of *defragmentation* all involve some measure of "emptying out" whatever is *not "Self."* When we are running *Cathartic* or *Kenostic* techniques we are always "emptying out" the emotional charges stored within *imprints* or the *thought-formed "beliefs"* we feed our energy into. In order to reclaim the energy, the ZU, the *Knowingness*, and the freedom of the *Spirit* that takes up this Human Condition, it is obviously required that we "purge" everything that is not the *Alpha Self*. These are all methods to ultimately return the *Awareness* of the *Seeker* back toward the "spiritual." The only reason there is any sense of "wonder" and "discovery" attached to this is because we have apparently forgotten that we have trod upon this very same *Pathway* at least once before, on our descent into the Human Condition for at least this present lifetime—and however many other countless times our *Spirit* has crossed with the planet Earth.

Why do we call an individual on the *Pathway to Self-Honesty* a "*Seeker*"? It is because that individual has set out to recover something that has been lost or forgotten. Apparently our "sense of *Self*" was misplaced somewhere along the way as we took on these shells. But the very fact that we have descended this *Pathway* before is of tremendous benefit to us in recognizing the truth of our *Self*, the Reality around us and all of Existence. Because it is this inner sense that allows us to recognize it when we achieve it or find it. It is that very recognition—that sense when reading or working through

Systemology that we somehow *already know* this information and find it strange that it should have taken so long for someone to systematize the knowledge—it is that very recognition that validates the Reality of this *Path* for the *Seeker*.

Make no mistake: although *Kenostic Processing* is introduced within the framework of solitary *Self-Processing*, the formula of "AR-SP-2" is a very effective standard *Pilot* procedure. In fact, although education and tools for directly *resurfacing imprints* are necessary to maintain certainty and confidence required for *Piloting* others, these other *Kenostic* methods of "analytical recall" are actually preferred—whether working alone or otherwise. Therefore, it may be stated at this juncture of research and discovery that "AR-SP-2" is the Route preferred when it may be applied; and that "RR-SP-1" is used only when necessary—such as when an *imprint* is stimulated. It may also be the case that an individual has kept *Awareness* so deeply wrapped up in their *imprints* and *conditioning* that the more intensive Route is necessarily taken only to free an individual up enough to actually achieve true *analytical* levels of *Awareness*. It is demonstrated that at the highest levels of *beta-Awareness*, a *Seeker* may effectively demonstrate greater *Self-control* in handling their own *imprints*.

I dispensed some details concerning the "RCC" *Reactive Control Center* and "MCC" *Master Control Center*[68] in regards to the

---

68 **Master Control Center (MCC)** : a perfect computing device to the extent of the information received from "lower levels" of sensory experience/perception; the proactive communication system of the "*Mind*"; a relay point of active *Awareness* along the Identity's *ZU-line*, which is responsible for maintaining basic *Self-Honest Clarity* of *Knowingness* as a *seat of consciousness* between the *Alpha-Spirit* and the secondary "*Reactive Control Center*" of a *Lifeform* in *beta existence*; the Mind-center for an *Alpha-Spirit* to actualize cause in the *beta existence*; the analytical *Self-Determined* Mind-center of an *Alpha-Spirit used* to project *Will* toward the genetic body; the point of contact between *Spiritual Systems* and the *beta existence*; presumably the "*Third Eye*" of a being connected directly to the *I-AM-Self*, which is responsible for *determining* Reality at any time; in *NexGen Systemology*, this is plotted at (4.0) on the continuity model of the *ZU-*

ZU-line and *Standard Model* for the *"Tablets of Destiny"* book. These centers of energy regulation are at the relative points of (2.0) and (4.0) on the *Awareness Scale*. At the level of "Thought"—ranging between (2.0) and (4.0)—we have various degrees of analytical thought capable of being achieved as *beta-Awareness*. This is not a pipe-dream: "Full beta-Awareness" is simply "Full beta-Awareness." That's just about bringing us back up to "zero," right? I mean, how is it that we have come to such a state in the Human Condition that this idea of being "fully present" in *beta* as our *Self* is such an amazingly extraordinary thing to achieve? We won't even try to unravel that with the time allotted to us today; suffice it to say, there is a range of potential *Awareness* while occupying the Human Condition, and as soon as our *Awareness* drops to a point of any degree of autonomy—at (2.0) or below—we are already down to only "25%" of our "Full" *beta-Awareness* as *Self*.

Now, I should point out here that our maths on this are not out of whack. Someone asked me about this the other day. If you look at the *Awareness Actualization* percentages for *Beta* range—up to (4.0)—there would seem to be something happening here that doesn't follow normal math logic—but that is only because of another key to understanding the *ZU-line* on the *Standard Model* that has not been sufficiently described until now.

Our understanding and graphic representation of the *ZU-line* is as a waveform or frequency. The vibration type increases as we move "higher"—away from the direction of the Physical Universe (KI) toward the direction of the Spiritual Universe (AN). Just so you know, the entire line represents an expression of *constant energy* between "Infinity and Infinity"—but we are mainly treating the positive scale of potential *is-ness* and *Beingness* between physical continuity[69] (0.0) and *Infinity*,

---

line.
69 **continuity** : being a continuous whole; a complete whole or "total round of."

which is plotted at (8.0) on the *Standard Model* or *ZU-line*. As we move up the scale and increase in frequency, each area or level actually demonstrates more energy in a space—more potential—but exceptionally more energy in the same relative space or distance of time.

Further advanced *Systemology* texts will be developed in the future concerning Universes—because this is actually something that *Alpha Spirits* have a causal responsibility of creating once actualized—but the idea of frequency and vibration is something that we are dealing with every step of the way. Right now, it is important to mainly understand one key aspect: the energy contained between, for example, the zone or range between (1.0) and (2.0) is far smaller, more finite, lower energy, less space in time, slower vibrations—and so forth—than the distance between (2.0) and (3.0); and so on up the *Scale, Chart* or *Model*. There are many ways in which this is directly demonstrable in *beta existence*. For example: an individual operating in a higher state of *Awareness* is able to conduct a greater amount of energy in space within a fixed amount of time—which is why they maintain an increased "reach" or greater sphere of influence$^{\infty}$ as a result—and they get more accomplished in the same amount of time. That's all that make's time relative; the frequency.

△ △ △ △ △ △ △

The light *Kenostic Processing* methods delivered in our "self-help" book are just as effectively performed as *piloted* techniques. Although we are focusing our attentions on the "analytical" levels of *Thought-frequency*, this *Tech* is going to contact emotional energy held in low frequency bands on *emotional encoding* and *conditioning* just like *Cathartic Processing*. This new *Tech* is quite effective in doing that—but most importantly it is effective in progressing *Seekers* on the

---

∞ The "Spheres of Influence" are treated later in the present volume.

target goal of this Grade,* which is to restore the personal ZU-energy vitality or *Awareness* of the Human Condition. These "analytical recall" methods provide a direct increase in conscious *Awareness*, use of the faculties inherent in our *beta-existence*, such as perception or memory recall. These attributes directly affect our demonstration of intelligence, and by this I mean not only "analytical" intelligence but the increased *beta*-perceptiveness that accompanies "emotional intelligence."

A directed use—either solitary or *Piloted*—of "AR-SP-2" should always emphasize the greatest "total recall" of *facets* to generate the most complete *resurfacing* of energy possible. In most cases, the suggested memories for contact during *Self-Processing* are not meant to directly stimulate deep emotional turbulence. Such happens only as a result of an individual's personal *programming*. If the general idea of taking care of personal possessions or being able to express one's imagination to others is restimulating a deep *emotional imprint*, yeah that is going to have to be dealt with—which is why we don't start people onto *this* ("analytical") Route without first providing emotional education, such as became a major focus of our former "RR-SP-1" instruction in *"Tablets of Destiny."*

Systematic *Self-Processing* may be performed with the same effectiveness as experienced *Piloted* processing so long as the *Seeker* is capable of maintaining the same strength, integrity and education level as a successful *Pilot*. This provides one additional reason why we simultaneously conduct these "Extended Courses" aimed primarily at *Pilot Training* for the "ISS Flight School," while at the same time we focus on the education, general materials and personal processing directly. It is important for me to point this out as public attention toward our methods increases, because there is one fundamental truth I want to keep in mind as we move forward in *Systemology*:

---

* *"Grade III—Mardukite Systemology."*

A tradition fails when it no longer duplicates the effective results of its originator. I do not wish to see a repeat of history with something carrying such great potential and futurist orientation as Our Cause: *Systemology*.

Increased education not only increases a *Seeker's* understanding of *NexGen Systemology* or even *Mardukite Zuism*, but it increases the personal effectiveness of *Tech* when applied to *Self* or others. And it is equally important. I have always found it interesting that a person, simply by learning to effectively use the Tech can assist in producing progress for another on the *Pathway* that they, themselves, have perhaps not fully actualized. That's kind of interesting to me. Perhaps it has something to do with the combined *Awareness* of the two people involved. But, since this is the case, it should also mean that we can all work together to more efficiently and expediently raise the *Awareness* of ourselves and each other by simply working together a bit. What a concept! We could, in a relatively short time, theoretically elevate the entire consciousness of the planet in this way. We *really* could.

Where we have previously targeted hidden and suppressed energy stores using intensive *"Resurface and Reduce"* methods,‡ these other light *"Analytical Recall"* methods—that lend themselves perfectly to *Self-Processing*—focus on what we general consider "conscious events" stored in our memory. I will not stand here and tell you that there is not a possibility that these methods will stimulate suppressed memories once the various *facets* are brought to full conscious *Awareness*, but as a person increases in their *Awareness*—which we now demonstrate systematically with a gradient scale—the ability to effectively manage personal memory and energy stores also increases.

It is only when an individual is brought out of a state of total *Awareness* that their *fragmentation* has any effect on *Self* or its perceptions. It is not the idea of *imprints* or *thought-forms*

---

‡ *"RR-SP-1."*

themselves that is the problem. Without any of these, we would probably not have a universe. But!—are *these* creations of ours, our own? Are they always in our control? Are we unknowingly the effect of our own cause?—or, are we totally *Self-directed*? That—*that!*—is what we are really concerned with. Because otherwise—an *Alpha Spirit* is entirely capable of realizing anything it wants to in existence without becoming imprisoned by it. But that requires freeing a *Spirit's* own actualized ability to do this—of which our channels[70] are repeatedly cut off the more we decide to identify with the Physical Universe (KI) and these bodies within it.

We must ask ourselves the same questions of these bodies as we do of the *imprints* they store and the *thought-forms* we interact with as a result. Because it's become exceedingly evident, the further we trek down this *Path*, that the bodies we are animating and the universe they exist in—*Alpha-Spirits* once systematized and *thought-formed* into *Being* in the first place. It's important to take moments to look around with each elevated point of *Awareness* once and a while—because you might just learn something more about yourself. We maintain environments as a direct product of ourselves. You want to see what is going on in a person's head?—take a look at how they manage their environment and what they are certain of in their responsibilities. The *Beta-Awareness-Test* is practically unforgiving on this one point alone.

Systemology Procedure "AR-SP-2" is intended to raise an individual's general *beta-Awareness* level just as is indicated on the BAT test and evaluations of the same. This automatically generates conditions for a *Seeker* that promote increased certainty in personal *Life-Management*—and at the basic core, isn't that what everyone is here for anyways? Why else do we practice any of the traditions, methodologies or techniques that have been provided out there? Don't be shy—each and every one of you has undoubtedly made an attempt to increase this in your lifetime by some type of "meditation" or

---

70 **channel** : a specific stream, course, direction or route.

"prayer" or "spell" or "affirmation"—a million ways of "auto-suggesting"[71] some temporary state that will simply help us better manage our lives—even if only by adding yet another layer of "belief" onto the pile that seems to get us through *just one more* day.

These old methods—sub-standard methods—obviously get to be very cumbersome. I say the word "obviously" because it is clear, at least to myself, that a lot of this is not working for us over here in the Western[72] world—or at least not effectively fast enough to provide the real results we need to see. I've never been to Tibet or Asia or China or Japan—I don't know how well the Eastern[73] methods are working for them over there, first hand anyway. I don't hear very good things about the state of anywhere on this planet right now, though. All I do know is that over here in Western civilization—it evolved from a genetic memory that extends from where Eurasia meets and all across to the Americas over several thousand years; we are playing a different *Game* over here. I suppose it might be easy for someone sitting up there in the hillsides, living in their mud-hut, to come tell us: "not to be a slave to our bills, *man*." Well, that's all fine and good, but we aren't going to be successful at playing the *Western Game* if we reject it, or try to attack it from the direction of "counter-belief." We don't need a revolt in order to upgrade an evolution to *Homo Novus*. We have reached such high faculties now as the Human Condition, that a *defragmented* individual has but only the need to *decide* to *be* more than Human.

---

71 **auto-suggestion (self-hypnosis)** : auto-conditioning; self-programming; delivering directed affirmations or statements repeatedly to *Self* in order to condition a change in behavior or beliefs; any *Self-directed* technique intended to generate a specific "*post-hypnotic suggestion.*"
72 **Western civilization** : the modern history, culture, ideals, values and technology, particularly of Europe and North America as distinguished by growing urbanization and industrialization.
73 **Eastern civilization** : the evolution of the *Ancient Mystery School* east of its origins, primarily the Asian continent, or what is archaically referred to as "oriental."

And that is all we are doing when *piloting* our *Seekers*—is freeing up their ability to decide and for the first time in perhaps thousands of years, returning the power of *clear choice* to the individual.

△ △ △ △ △ △ △

Basic *defragmentation* techniques explored in *Systemology* are aimed at freeing up the personal energy that is tied up, or wound up, in *imprints* and *thought-forms*—and generally it is only wound up in those which we have lost control of, or responsibility of, and therefore are not being treated as our own. In *Self-Processing* methods, we are using *analytical* faculties to reduce the erroneously cemented personal stores of *ZU*. This *ZU* is nothing more or less than the essence of *Life* itself, but we are treating the parts of it we can experience or measure—which for our purposes is *Awareness*.

We introduce the *Standard Model* and *ZU-line* and *Awareness Charts* and *Scales* prior to instructing the techniques of "AR-SP-2" for a good reason: they all provide a means of orientating our concept of *Self* in existence, for which we have few other valid landmarks or benchmarks or signposts on which to evaluate. If you can imagine the orientation of *Self* in a space and time without measure, one can imagine just how difficult it might be to chart any definitive "pathway" of movement within it. Seriously—an "I" floating out in the wide open spaces has no concept of direction until it *creates* one, fixes upon one, and then orients all other certainty and knowledge to it. Well, are you beginning to see? If not, you will by the end of this lecture series and its "book."

Alright—so you need some familiarity with the *Awareness Scale* if you expect to meter your results or really gauge anything during a process. When we say to "use a process so long as it is effecting change," we are referring to the *Awareness Scale*—or one of its derivatives—as a gauge. A *Seeker* can expect that when they hit an erroneous solid on the *ZU-line* that they are going to dip down into levels of emotion and fragmented thought in order to first bring memories up to the surface and onto the screen for our analysis. In the former *Cathartic Processing* methods, the deepest emotional turbulence, which is generally nonsensical other than pain and enforcement, is simply burned off through repetition, but it is not necessarily brought to a scrutiny up at the analytical levels; it's just being burned off as fuel—fuel to transmute or transform the *imprint* or *belief* into free flowing ZU potential again. We aren't stripping away anything that is the "true" *Alpha Self*—we are simply shedding these layers of skin that aren't *Self* and which seem to only keep us bound and wrapped in the convolution that is *beta-existence*. The irony is that when you get into higher levels of *Awareness*, none of this stuff that seems confusing and mysterious to the common man is actually complicated. It is only in the face of uncertainty that we find convolution and fragmentation. Such distortions are not even on our radar once we can maintain *Awareness* in the clear light of Truth.

A *Seeker* is only able to experience realizations at their level of *Awareness* or below it. This means that the experiences resurfaced directly in "RR-SP-1" are only effectively *analyzed* when a *Seeker* is above the level of which those *imprints* are wound up, fixed or cemented. For example, a person who is afraid—suspended somewhere around (1.5) on the *Emotimeter* with approximately 15% actualized *beta-Awareness*—is able to easily recall and revisit experiences scaled as "fear" or lower on the *Scale*. But, they won't be able to process anything rationally in the range of Thought above the mechanisms of the "RCC" *Reactive Control Center* at (2.0).

Therefore, these processes, although revisiting and resurfacing memories which may bring a person temporarily below this threshold while experiencing them, may then be able to carry the energy back up to *Analytical* degrees of *Awareness* and discharge it there using higher vibrational Thought. This is partially what makes "AR-SP-2" a new effective advancement with greater range of application than knowing and using "RR-SP-1" in exclusion with no other considerations. You want to be running something until elated, extroverted and enthusiastic about the future of your lifetime and potentials *in spite* of the memory that such and such has happened. So, you run it through, experience it, analyze every *facet* and then decide if there is anything still worth adding to the files. But it should be analyzed and not simply piled up.

It is important that we *consider* all of our past *considerations*[74] while on the *Pathway to Self-Honesty*. This would be one of the key reasons for the effectiveness of these basic procedures. It is important for us to take these things out and look at them, otherwise we are simply storing them up and carrying them around with us arbitrarily. And they *do* affect us—*especially* when they are hidden away, folded up and put away in our pocket. Yeah—that's a sure way *not* to be rid of something.

It is important to realize that we are on a mission to recover our true potential—not add something that wasn't already there. That is a big misconception about mysticism and magic and spirituality. But a truth is attained only on the *Pathway to xSelf-Honesty*, otherwise we are simply practicing Self-hypnosis and adding another layer of programming without sorting out all that lies beneath. We put considerable energy into telling ourselves that things are not as they are, instead of analyzing the original belief we agreed to in making things as they are. And this is one of the reasons why our personal ZU *Lifeforce* and energy has the appearance of being depleted over time—with age and experience—although the actual *Spirit* or *Self* is no different—*Self* is not changed; meaning that the actual energy of *Life* that was always supplied as a constant, is no different at the end of one's present physical lifetime than at the beginning. Something else has changed—but it is not the energy being *fed into* the system.

---

74 **consideration** : careful analytical reflection of all aspects; deliberation; evaluation of facts and importance of certain facts; thorough examination of all aspects related to, or important for, making a decision; the analysis of consequences and estimation of significance when making decisions.

Over time and with more accumulation of experience, *emotionally encoded imprints* and libraries worth of *thought-formed beliefs* are as like stacks of important books that we have agreed to, taken responsibility for, even dabbled with a bit—but then never really get around to "owning" the knowledge of for ourselves, so they pile up like walls and barriers to seeing anything more—because you see, we have already agreed to the *Reality* of these stacks once. They aren't going to just *go away* on their own. You can try wishful thinking—most of you probably have already. And if we are going to pretend they aren't there and just let them pile up while maintaining some alternate hypnotic state about it otherwise—well, I tell ya, that's the basis for a *psychosis* right there. Pure and simple.

By bringing the efforts of our past to the surface—including those efforts of others toward us, we are able to bring the *moves* and *counter-moves* of this *Game* to a scrutiny. This is the only way we could possibly earn any knowledge or actual information from our experience. Otherwise, experience is a rather *fragmenting* aspect of *Life* with no real use. If its purpose is so that we can learn, than we must be bring it out and learn—thus being *Aware* or increasing our *Awareness* as a result, rather than losing our *Awareness* to personal databases and libraries chock-full of all the *Lifeforce* we have chosen to file away irreverently. Basic *Systemology Processing* and *Self-Processing* is applied to change this. It puts the *Seeker* on a track where they are able to unravel and free up all of the vitality that is already theirs to begin with—just hidden away and forgotten. It is about time that we *remind* people just how beautiful and amazing the true *Alpha Spirit*—the *Self*—really is; how beautiful and amazing *Life* can *really* be.

∆ ∆ ∆ ∆ ∆ ∆

The procedure outline or formula for basic *Systemology Processing* was released in the book and conference for "*Tablets of Destiny.*" Therein, I described a seven-step procedure as defined by a list of ancient Sumerian cuneiform "signs"[75] representing words, concepts or phrases. Those steps described do not *only* apply to "RR-SP-1" exclusively. There is a basic underlying formula at work with that one—very powerful and effective—clearly devised by priests and priestesses of the ancient Temples, which may be applied to further forms of *Piloted Processing*. Obviously we are introducing the premise[76] of "AR-SP-2" for purposes of *Self-Processing*, but as I said before: this method is effective for *Piloted* or solitary use.

Systemology Process "AR-SP-2" may be conducted as a basic two-step process of *Self-Processing* while using the "self-help" book—or the steps may be used to modify the instructions for *Piloting* given as "RR-SP-1" in "*Tablets of Destiny*." If used as a *Piloted* procedure, the two steps from "AR-SP-2" would be effectively replacing the "step three" and "step four" of "RR-SP-1"—meaning that instead of the previous steps aimed directly at "resurfacing and reducing" the *emotionally encoded imprints*, we are replacing with instructions for "analytical recall" in this updated *Tech* known as "AR-SP-2."

The two steps, as defined by Sumerian cuneiform, are:

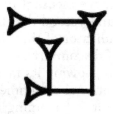

3. **SI** — "to recall; remember; be conscious of in Mind."

---

75 **cuneiform signs** : the cuneiform script, as used in ancient Mesopotamia, is not represented in a linear alphabet of "letters," but by a systematic use of basic word "signs" that are combined to form more complex word "signs."

76 **premise** : a basis or statement of fact from which conclusions are drawn.

4. *SUG* — "to empty out; to clear; strip away; make naked or bare."

Many former spiritual leaders have referred to *Kenostic* methods as, very literally, an "emptying out." And I am not about to say that this is inaccurate, but when we think of the similarities between *Cathartic* and *Kenostic* methods, both seem to involve a basic "emptying" or "discharge" or "clearing" of some energetic restriction that had once diminished the *Awareness* maintained as *Self*. In order to provide some greater sense of distinction, the *Seeker* should consider the "Analytical Route" to be a systematized process of "releasing the hold" *of-and-from* such and such.

In analytical processing work, there is an emphasis on recall and analysis of *facets* even more than the actual *subject* of the incidents and events themselves. It is possible, as always, that in the process of scanning various personal experience that some are going to be more pleasant or unpleasant than others—and it should be understood that the first "response" or "reaction" is a good indicator of this "hold" that needs to be "released." Any type of discomfort or emotional reaction to a *memory* is *fragmentation* showing its face—entirely dependent on some degree of emotional attachment and reinforcement as Reality of whatever it is becoming a *cause* to our *effect*—which should only ever be *Self*. By recalling as many of the *facets* at play in our memory of any event we are releasing the energetic stores that are tying that superfluous data up beneath the surface of our conscious memory—and actualized conscious is really the only place that memory serves any analytical function. Therefore there is not even a logical reason to keep energy bound up in *imprints* and *beliefs* that we aren't even making conscious use of, but of which affects us—

and we seem to have limited space or resources to maintain these *solids* during a single lifetime in *beta-existence*. Once an organism has accumulated sufficient non-survival *imprinting*, it will begin to die.

As a *Seeker* progresses on the *Pathway to Self-Honesty*, the level of *Self-Actualization* is directly aligned with the degree of certainty maintained concerning *Self* and its causal role in the universe.

> It is when we are at *cause* that we are most *actualized*. It is when we are surrounded by *effect* that our *Awareness* is lowered.

I'm not sure if I can even put that in simpler terms. Effective personal management, such as measured on the BAT test, is a *variant*—it differs from person to person—based on the actualized personal level of *beta-Awareness*: hence the purpose of the test as a basic measure. It's only a guide—or a gauge—used to measure progress while using *Cathartic* and *Kenostic* processes. The *Emotimeter* may be used as a basic guide to *resurfacing* and *recalling* energies and being sure that we elevate their energies to the highest degree possible. In doing so, the *Seeker* is systematically reestablishing their basic *Identity*—which naturally carries an increased *Awareness*, extending up even to *Alpha* levels as you may have guessed.

Effective "*Analytical Recall*" requires identifying and analyzing any *facets* associated with recalled events and instances. Only after we are able to bring our *Life*-experience to a scrutiny; analyzing the information; evaluating validity or truth; evaluating the effectiveness or rightness of our beliefs; and everything else that we have accepted or agreed to as Reality —only after all of this can we say with any certainty that we have yielded some kind of knowledge, something real.

> Even if we do, in the end, determine that *Life*-experience accumulated in the Physical Universe *is* actually mostly erroneous, well, then I guess we still have learned something real.

And the recall of events must be firmly rooted in our actual memory as we believe things are or have experienced them—not simply a result of what we are told things are or how things were for us. When we take a good look at our memory that we keep all balled up in a corner—when we take it out and look at it once and for all, we clear out the clutter and find what, if anything, there is to appreciate about the experiences we attach our beta-personalities to so strongly. Because that is what we do. So, let's do everything we can do to clear out the clutter from this world—because its getting pretty murky. So, let's start clearing it all up. And it starts with *Self*. Thank you.

UNIT THREE
SELF AWARENESS

RAISING LEVELS OF PERSONAL AWARENESS

The pursuit of *"Self-Actualization"* or *"Self-hood"*—the perfection of *beta-Awareness* during this lifetime—is a natural pursuit of all *healthy* lifeforms. All individuals are inherently working toward the continuation of their existence as a *being* —and the uppermost optimum levels that this may be experienced as Reality. Most individuals carry an intuitive knowledge that there is much more happening beneath the surface of *Life* and *events* than we have been led to believe. And naturally we are correct in this assumption,[77] because we find evidence for it each time we are able to exceed the "norms" of the Human Condition.

Regardless of however it is that we have come to be in this state—we are *here* and we are all *Seekers* for the truth that will unlock and unfold the *Self* that is buried beneath the *imprints* and *programming*. The *Pathway to Self-Honesty* is largely a route back to our own *Self*, the true and basic *Self* that is *Eternal* and maintains an *Alpha* spiritual existence in a Spiritual Universe (AN). It would seem strange that this should be something sought to *Be*—that we should somehow have misplaced the total *Awareness* of *Self*, but that is what happened. This will become increasingly "Self-evident" as you spend more time systematically "recalling and analyzing" *fragmented programming* that you have at some point agreed to and now take for granted. All of the accumulated conditioning and programming—however erroneous it may be—has never stopped operating its influence on our Reality. That is up to us to *squash*—just as it was up to us to take it on in the first place. We are creatures very much disposed to *having* "things" even when they do not particularly lend well to our survival.

The highest ideal states of the Human Condition are its ultimate *Destiny*—the destination by which we will arrive if we apply enough "right thought" to "right action."[78] The actual

---

77 **assumption** : taking or gather to one's Self; taking possession of.
78 **destiny** : what is set down, made firm, standard, or stands fixed as a

gauge of this "rightness" is determinable by the level of *Awareness* and *Self-Honesty* achieved. This is what results in "right experience" or "true wisdom"—as the word "true" replaces "right" in semantics of some philosophical schools. One of the reasons *NexGen Systemology* is developed for the 21st century is because of this very semantic issue: where so many different religious and philosophical schools have each spread the *fabric* of debate to such sheer thinness by alleged pursuits of truth; it has been torn to bits with each overstretching their portion in some opposing direction by adopting contrary or obscure terminology. Most often there are no philosophical issues—only semantic ones. And since the Mind-Systems and even the RCC are very often dealing with "language" for programming, this matter of "semantics" is no arbitrary thing. This *too* can be "analyzed" during systematic recall.

The most basic state of *Self-Honesty* achievable during this lifetime equates to the total *actualization* and *realization* of *Self* —the "*I*"—as present to control the *beta/genetic body* free of fragmentation: meaning reactive-responsive conditioning and erroneous beliefs about what *Self* or "*I*" *is*. Once full *Self-determination* is resumed, the *Seeker* achieves *Self-Honesty* in their ability to be *Self-directed*—and they may then begin to work toward higher ideals: such as the nature of *beliefs* and *creations* they may now knowingly create and experiment with without becoming their effect, or enslaved by these *thought-formed* creations. All creations and manifestations[79] are some degree of *thought-formed* energetic creations; the ones that we say are more "solid" are simply agreed to by more people to a higher degree of reinforced intensity.

Delivery of effective *NexGen Systemology* education and *processing* is a combination or alternation of focus between

---

constant end; the absolute *destination* regardless of whatever course is traveled; in *NexGen Systemology*, the "*destiny*" of the "*Human Spirit*" (or "*Alpha Spirit*") is infinite existence—"*Immortality.*"
79 **manifestation** : something brought into existence.

introversion and *extroversion*. First, the *Seeker* employs basic *Tech* that will draw them inward toward *Self-Analysis* and *Self-Evaluation* to scan and survey what is accumulated in emotional *stores* and intellectual *databases*. Once these *facets* are brought up to a range of full analysis, the discharge of their energy allows for an increased *Awareness* or elation that is conducive to an increased receptivity to higher level education and understanding. This is how "right knowledge" and "right experience" results in the "true wisdom" that is consistently sought but seldom found when following traditional well-beaten paths. The real *Pathway to Self-Honesty* is a retracing of the routes that only one individual can be absolutely certain of—the *Seeker* that is traveling it for themselves. Too many overt attempts at distorting traditional paths have been found for us to ever rely on them again if we are to bring the Human Condition of *Homo Sapiens* toward its next great evolution as *Homo Novus*.

The *Pathway to Self-Honesty* is a natural evolutionary "unfoldment" of *Self*—marked by a return to our most natural spiritual state and realized in *this* lifetime—in *this* "beta" existence. Just as every person alive thinks the world to be in line with their beliefs—and that they are in need of no more "sense" or "reasoning" than they already possess—so it is with the person who has become so hypnotized by the experience of a Human Condition that they are incapable of seeing that there is any greater "*Awareness*" to be had in *present time* than they are already maintaining. The amount of programmed reasoning and "reactive-response" conditioning maintained will automatically justify and find cause for any aspect of existence to the extent of the knowledge maintained.

Whatever we hold as truth will be used to validate everything else that we receive information or experience from—and thereby it will only serve to reinforce the originally established belief, premise, *solid* or *postulate*.[80] Without systematic

---

80 **postulate** : to put forward as truth; to suggest or assume an existence

review of these *formations*, we are left to a diminished experience of *Reality* based on preexisting *filters* to that experience. As more *filters* are placed in the path, the clarity and luminescence of *Clear Light* is greatly dimmed—until we are holding so little of this original source *Awareness* that we succumb[81] into material demise. This can be delayed—or in essence—avoided altogether when we set our sights on *Infinity*. We increase *beta-Awareness* to maintain highest optimum existence during this entire lifetime—promoting our abilities to formulate necessary spiritual development experiences required to fully and consciously direct our *Alpha Spirit* in the same way we first learn to fully actualize *beta-Awareness*.

> By actualizing the *Self* on the *Pathway* in *this lifetime*, we live longer healthier lives and we do not deteriorate materially from a reduction of *Awareness*; we merely shed the physical body at some distant point when it is deemed no longer necessary for us to use in our actualization of a *Self-directed* continued spiritual evolution *exterior* to the Physical Universe (KI).

Actualized ability and realization of *Self-direction* must always be *regained*; it is no more automatic in the Spiritual Universe (AN) than in the Physical, contrary to what many may believe about some instantaneous perfection achieved—for with that idea of perfection would come complete and total dissolution of *Self* into *One Beingness* which is *Infinity* (8.0). Therefore, we *do* maintain a basic or primary "I" personality as the *Alpha Spirit* (7.0) or "Self" in between spiritual lifetimes—and these lifetimes may not be restricted to the planet Earth (such as with the *genetic body*) although at this juncture of the physical timeline,[82] this planet is the primary center for current spiritual activity in the solar system.

---

to be; to provide a basis of reasoning and belief; a basic theory accepted as fact.

81 **succumb** : to give way, or give in to, a relatively stronger superior force.

82 **timeline** : plotting out history in a linear (line) modal to indicated instances (experiences) or demonstrate changes in state (space) as measured over time.

These *genetic vehicles* are organic bodies best engineered for sustaining an existence on *this* planet specifically—and even then only within a very marginal range of acceptable conditions. It is possible that we are preparing for alternatives for physical survival—just as the Human Condition was once selectively evolved and implanted with programming to maintain (or possibly enslave) sentient spiritual *Life* or *ZU* on this planet, for whatever the purposes of these original Reality Engineers. There are many ancient sources—especially the cuneiform tablets* of Mesopotamia—that lend us clues about these details for those interested in the *historical records* pertaining to "*Mardukite Systemology.*" But for our present purposes, we will focus specifically on the information and education that lends itself to advance the *Seeker's* developments using *Self-Processing*.

FIRST STEPS WITH "SELF-PROCESSING"

This introductory information for *Self-Processing* pertains directly to the management of our responsibilities regarding various conditions within our most immediate spheres of influence and responsibility:

a.) our physical possessions—including the "body";

b.) our basic management of emotion; and

c.) our basic management of thought activity.

Over the course of *Self-Processing*, you should accept resurfacing of any "reactive-responses" that occur or erupt in connection to any memory or concept being conjured to Mind. If you wish to apply *cathartic processing*, you can "run these out" and exhaust them by reduction that way, but this is no longer the *only* Route provided to *Seeker*. The alternative is to bring the full energy up to the realm of conscious *Aware-*

---

* For example, the collection of cuneiform tablet source material compiled within "*The Complete Anunnaki Bible*" edited by Joshua Free.

ness and simply *confront* it—or *dissolve* it—with greater frequency energy. Some individuals do a light version of this naturally: "going over what has happened to them, over and over again to desensitize the negativity," then usually "going back to happier times" of memory whenever faced with difficulties or stress. In doing so, a person is actually raising their level of *Awareness* and personal efficiency, drawing out from higher frequency stores of ZU—even if temporarily—in order to properly combat, confront or balance counter-efforts in a situation. This is actually natural, yet many have lost *Self-determinism* and consistent "drive" to *be* their own reason; their own *cause*—and this is when an individual begins to lose control of the mental faculties necessary to properly direct *Self* in efficient and creative management of *beta-existence*.

Throughout the developmental work in *Systemology* we have discovered so many tenets and fundamentals that seem as though they should be "common sense" evenly distributed throughout the Human Condition, yet if they are, they remain beneath the surface. It is a "recognition" of these premises as "true" to us that offers us any clues that we are on the right track. But then, why hasn't this already been done before? Well—it *has*, but in the past it was only realized at *that* level of understanding or *realization* and no further. At each level of development for the Human Condition, the systems enlarge themselves to conceal the truths in more and more layers of *fragmentation*. So, at each point in the work—even when not directed specifically in the text—

- If you find yourself reaching new elated realizations or something rings loudly because somewhere deeply laden *"you already knew this,"* take a short session to *Self-Process*:

  —RECALL the *time* you put the *Alpha* belief aside or displaced it for another person's *Beta* belief.

  —RECALL every perceived *facet* of that event-experience until it is *analytical*.

  —ANALYZE the *facets* and *total experience* intellectually until you are high on the *Emotimeter*.

- If you find difficulties with the above or you find that the *encoding* or *programming* of a particular event-experience is not easily discharging ("flattening" or "dissolving") with *Self-Processing*, shift your focus of the session toward raising personal ZU-frequency another way:

—RECALL a *time* when personal experience and certainty resulted in success.

—ANALYZE the *facets* experienced with that success until you are high on the *Emotimeter*.

The more we are certain about anything we manage, or have similarly managed before, the more energy we are able to apply to it. This might seem like it should just naturally *be* the case—yet the reasoning behind it has always been clouded in fancy words and obscure ideas about "*consciousness.*"[83]

An individual will always "think" they are applying full attention and full *Awareness* to their management of *Life*, because the *Self* does not know any better—unless it is "allowed" to know better from its *beta-experience*. If those channels are blocked by emotional turmoil, enforced realities and erroneous beliefs, this becomes the totality of *Awareness*, which is still a "totality" *of* what is capable at any given time, but certainly not *all* of what is capable *in* totality. To put it another way: this drinking glass will always seem "full"—but greater *Awareness* provides a larger glass. And since totality of our potential is a very large glass, there are some who intuitively seem to be *Aware* that they are missing something from their "normal" experience here and they go looking—*seeking*. Since everything that we actually "*are*" is unchanging, and simply buried, it would seem logical that the first steps on the *Pathway of Self-Honesty* and *Total Self-Actualization* should begin with basic "*Self-Processing.*"

---

83 **consciousness** : the energetic flow of *Awareness*; the Principle System of *Awareness* that is spiritual in nature, which demonstrates potential interaction with all degrees of the Physical Universe; the *Beingness* component of our existence in *Spirit*; the Principle System of *Awareness* as *Spirit* that directs action in the Mind-System.

Even if you are not working with an experienced *Systemology Pilot*, it may sometimes be helpful to have someone else—a friend or partner perhaps—reading or "inputting" the "*processing commands*"[84] of a session in order to keep your *Awareness* fully on the session "in phase"[85] and not concerning yourself with the dictation of instructions. This phenomenon of ritually shifting "out of phase" was first discovered during our mystical experiments where we learned quickly that an operator too concerned about words and operations of ritual-ceremonies actually rendered its effects useless. If you are working entirely alone, *Self-Processing* is just as effective so long as the attention is able to be securely *Self-directed*. It may be found helpful to use a ruler, or cut out a rectangle from an index card, to concentrate on one specific "*processing command*" at a given time—such as those lines given above at the beginning of this section on "First Steps"—without drifting to scan the page above and below it. We have even experimented with these procedures at the Office by writing a single command on an index card, reading it and then assuming the direction of the command from Self and not as an instruction to Self.

"SELF-PROCESSING" LANGUAGE AND COMMUNICATION

All vibrations, frequencies, energetic interactions and even the very commands used in *Systemology Processing* are forms of "communication relay." We tend to think of communication only in terms of audible language—and this is one reason

---

84 **processing command line (PCL)** or **command line** : a directed input; a specific command using highly selective language for *Systemology Processing*; a predetermined directive statement intended to focus concentrated attention.

85 **phase alignment** or **"in phase"** : to be in synch, in step or aligned properly with something else in order to increase the total strength value; in *NexGen Systemology*—referring to alignment of *Awareness* with a particular identity, space or time (such as being *in* Self *in* present *space* and *time*).

it is effective in *"processing-out"*⁸⁶ *fragmentation*; because in many ways it is the words and ideas represented in language that are used for *programming*. But words *are* symbols, just as much as any other esoteric representation, and they have no power in themselves except what we give them as *meaning*—the *"semantics"*⁸⁷ that they are representing. It is important not to become overly concerned with the words and language used to describe *Self-Processing* as long as the basic instructions are understood and carried out consistently.

Metaphysical and spiritual techniques frequently employ words and symbols to produce a result—but the result is entirely dependent on the individual practitioner and their own *processing* of language. We tend to respond quite readily to language, but we often have overlooked how much of our "response-reactions" are linked to that *programming*—and this is one reason why *"Analytical Recall"* is so effective: because it brings *facets* and associations out in the open that are held as part of the meanings, even if inaccurately so. We are usually never certain of this until we *evaluate* it *Self-Honestly*. Otherwise, we are left with taking many things for granted on "automatic processing" (at the level of the "RCC" *Reactive Control Center*)—*beliefs* and *imprints* we were not even *Aware* of because they are built upon premises and foundations of thought that we formerly agreed to. Even if only in casual acceptance or a stronger enforcement of *reality*, there is some point where we accepted that "A" is "A"⁸⁸ in order to share

---

86 **"process-out"** or **"flatten a wave"** : to reduce *emotional encoding* of an *imprint* to zero; to dissolve a *wave-form* or *thought-formed* "solid" such as a *"belief"*; to completely run a *process* to its end, thereby *flattening* any previously *"collapsed-waves"* or *fragmentation* that is obstructing the *clear channel* of *Self-Awareness*.

87 **semantics** : the *meaning* carried in *language* as the *truth* of a "thing" represented, A-for-A; the *effect* of language on *thought* activity in the Mind and physical behavior; language as *symbols* used to represent a concept, "thing" or "solid."

88 **A-for-A** : meaning that what we say, write, represent, think or symbolize is a direct and perfect reflection of the actual aspect or thing —that "A" is for, means and is equivalent to "A" and not "a" or "q" or "!"

Reality with our fellow operators of the Human Condition. Other beliefs and reality validations are then all systematically built from and arranged upon these former foundations.

Our physical, emotional and intellectual "mobility" is directly related to the free-flow of personal ZU—the *Spiritual Life Essence* that is our sense of *Knowingness* and *Beingness* as "*Self*." It is very important during "Analytical Recall" that you are able to *recall* and *analyze* your stores of memory as "actions." This is one of the perceivable *facets* of our information: "motion." It is as real and as apparent within *programming* as any other "sense" or variation thereof. All sensory information picked up from the environment and delivered to *Self* is also a form of energetic communication. We find that everywhere we turn regarding the interactions of *systems*, there is a communication taking place—and clearer communication of energy means a more efficient system. This is something we find as a general rule in *Systemology* without exception.

We have introduced the subject of *facets*[89] and *perceptions* within various *Systemology* texts. The specific examples provided below are a few concise "directions of attention" used to contact the programming information necessary to free most of the ZU cemented in their emotional storage without targeting exclusively "painful" experiences, such as with *Cathartic Processing*. In the present instance of *Self-Processing*, energy from the following *facets* should be *resurfaced* using "AR-SP-2" and carried through analytical ranges to true *Knowingness*.

LIGHT-WAVE FACETS (*Examples*)
- Scene/Places
- Brightness

---

89 **facets** : an aspect, an apparent phase; one of many faces of something; a cut surface on a gem or crystal; in *NexGen Systemology*—a single perception or aspect of a memory or "*Imprint*"; any one of many ways in which a memory is recorded; perceptions associated with a painful emotional (sensation) experience and "*imprinted*" onto a metaphoric lens through which to view future similar experiences.

• Symbols/Things

SOUND-WAVE FACETS (*Examples*)
  • Language/Speech
  • Loudness
  • Noise/Tone

MATTER-PARTICLES (*Examples*)
  • Touch/Feel
  • Hardness/Softness
  • Smells/Tastes

MATTER-WAVES (*Examples*)
  • Motions/Actions
  • Wind/Movement
  • Humidity/Temperature

It is easy to safely and effectively build personal skills using "*Analytical Recall*" by first targeting event-experiences that pertain to the type of *facets* you want to get a better "sense" of—and then advancing that skill with the full recall of all related *facets*. It is sometimes easier to start with memories that we are fond of and which we often conjure up in our Mind's Eye[90] for "personal strength" or some other shift in state as the need arises. As you increase skill with operating the Mind-Systems on command, this method of efficiency may even be carried to many other areas of one's personal *Life-Management*. This whole "systemology" of methods and education as *Tech* works together holistically—one merely needs to start working with it to see the results and to build upon any effective results realized. Take what works and consider whatever else for another time when perhaps you cycle

---

90 **Mind's Eye** : the activities of the "Third-Eye" (or actualized MCC) where the *Alpha-Spirit* directly interacts with the organic *genetic vehicle* in *beta-existence*; *Self-directed* activity on the plane of "mental consciousness" that is maintained between "spiritual consciousness" of the *Alpha-Spirit* and the "physical/emotional consciousness" of the *genetic vehicle*; the "consciousness activity" *Self-directed* by an actualized WILL.

through this work again and discover an even deeper meaning behind all of these carefully selected words. Practice with the following *processing commands*:

—RECALL a scene that was very colorful. *What was it? Where was it? What did you hear?*

—RECALL a scene that was very peaceful. *Where was it? How did it smell? What did you hear?*

—RECALL a sound that was very loud. *Where was it? What was loud? What else did you see?*

—RECALL a sound that was very soothing. *Where was it? Who was there? What else did you hear?*

—RECALL a sound that was not understood. *Where was it? What/Who made the sound? What emotion did you feel?*

—RECALL an object that felt sharp. *Where was it? Who else was there? How bright was the space?*

—RECALL an object that felt soft. *Where was it? What did you hear? What emotion did you feel?*

—RECALL a taste that was sweet. *What was it? Where were you? What did you hear? Who else was there?*

—RECALL a smell that was strong. *What was it? Where were you? What actions were happening?*

—RECALL an object moving away from you. *What was it? Where were you? What did you hear? What else did you see? Who else was there? What did you smell?*

—RECALL an object moving toward you. *What was it? Where were you? What did you hear? What else did you see? Who else was there? What did you smell?*

—RECALL the time you first heard of Systemology. *Where were you? What did you hear? Who else was there? What emotion did you feel?*

—RECALL the time you started performing this "*Analytical Recall*" session. *What position was your body in then? What position is your body in now?*

Whenever you complete a *Self-Processing* "session"—regardless of the results or how many processes have been run—the session should away be formally ended with analytical review of the session and a revisit to positive successful memories that an individual holds "certainty" of. Attention may then be extroverted toward an object or recreational activity to cement a positive experience. This also takes energy discharged from the session and stores it temporarily in our high-frequency banks. The more energy we can bring up to these higher frequency states the "lighter" we feel and the more "able" and "certain" we are in *Self-directing* personal management of Reality in everyday *Life*.

THE SPIRITUAL SIDE OF SYSTEMOLOGY PROCESSING

Systemology Processing is an "applied spiritual technology" that is effective for *Seekers* following any personal mystical, spiritual or religious foundation. It does adopt some use of ancient Sumerian and Babylonian cuneiform *terminology*, which is directly derived from a very ancient—if not the oldest—systematized Mesopotamian religious tradition widely recognized in Babylon as *"Mardukite Zuism."* Although a similar revival of *"Mardukite Zuism"* is in operation today from the present author, the methods presented as *"NexGen Systemology"* are only applied to that religious tradition as a *Spiritual Technology*, but are not in themselves exclusively "religious" practices. These *Systemology* practices essentially echo—and are based upon—"energy work" and "counseling services" provided by ancient priests and priestesses serving in the Mardukite Temples of Babylon. Other mystical practices derived from the same have also survived through various "secret societies" and "magical orders" throughout the ages, but few have been privileged the faculties now available to us in modern times to trek back directly to the distance source of the *"Ancient Mystery Tradition."*

For some *Seekers*, "religion" and "spirituality" are very *fragmented* subjects, laden with very heavy *conditioning* and other manners of enforced reality, guilt and so forth. Many are able to simply look past all of the semantics and recognize that these words simply carry a stigma as a vehicle or catalyst for *poor control* and *confusion*. This is one aspect that should be cleared up and "run out" in *Self-Processing*—especially if the emotional stores tied to these words are inhibiting progress on the *Pathway*. This is one of the reasons that the *Pathway* becomes a very "individualized" effort for the *Seeker*; because it is up to *you* to discern just how restricting, distorting or aberrated[91] any particular *imprint* or *belief* actually is.

—RECALL a time you were in a church. *Where is it? What do you see? What do you hear? What do you smell? What time of day is it?*

—RECALL an earlier time you were in a church. *Where is it? What do you see? What do you hear? What do you smell? What time of day is it?*

—RECALL a time you were made to go to a church. *Who is enforcing you to go? What are they saying? What else do you sense? What emotion do you feel?*

—RECALL a time you were made to feel guilty. *Where is it? Who else is there? What do you hear? How bright is the space? How large is the space?*

—RECALL the moment you decided to go some place on your own. *What position is your body in? What actions or motions are happening? What emotion do you feel making the decision?*

—RECALL the place you decided to go to. *Where is it? What do you see? What do you hear? Who else is there? What do you smell? How bright is the space?*

—RECALL a place you enjoy visiting. *Where is it? What do you see? What do you hear? Who else is there? What do you smell? How bright is the space? How large is the space?*

---

91 **aberration** : a deviation from, or distortion in, what is true or right.

If you can simply "change your mind" about a certain "aspect" as a result of improved education and simple *Self-evaluation* to gauge results, then it is possible that there is not a significant amount of *fragmented encoding* beneath the surface regarding *that* particular aspect. This is how we would prefer all of it to be. "Bad information" is simply "bad information." *That* can be remedied.

Fragmentation reflects not only an inability to recognize "bad information" *analytically* but also an inhibition to receive "right information" or demonstrate any directed interest in changing the *way things are*.

If an individual firmly rejects any "other" *spiritual* existence behind the Physical Universe (KI) and that we are *Spirits* "clothed" in the Human Condition—it would be very difficult to find reason for any effectiveness of *Systemology Processing*.

You will notice that instructions for basic *Self-Processing* are targeting the most available or easily recalled "conscious" memories, beliefs, thoughts and emotions related to event-experiences. All the command is calling for is a RECALL. It does not specify from "what time"—although the basic logic is to continue back to earlier and earlier incidents each time you cycle through the *process* that directly relate to the theme, call or command line. When *Self-Processing*, you should always begin with what is most readily available without much strain. As more of the *facets* are called up and analyzed, the energy attached to them may be strung or linked to previous events, sometimes ones that we have forgotten or misplaced *Awareness* of, although we were apparently "conscious" at the time.

Basic storage and retrieval of this information is appropriated to the realm of "analytical thought" and its databases. When maintained as *emotional encoding* that is *conditioned* beneath the "surface" of *Awareness*, it is both: withholding more of our free flowing energy; and not within our conscious memory stores for valid usage. We are not "erasing" memories—we are freeing up the resources in maintaining what we aren't even properly remembering or using. The very fact that we can effectively demonstrate two independent parts of the Mind-System—which we refer to as the "RCC" at (2.0) and "MCC" at (4.0)—demonstrates that there is certainly more at work behind our *beta-experience* of Reality than simply physical matter and hard-wired genetics. This acceptance of a *Self-Aware Spirit Self* as a fact—and the description of this unique spiritual energy of *Knowingness* and *Beingness* called "ZU" in the language of ancient Mesopotamians—is essentially the *only* spirto-religious connotation implied directly by *NexGen Systemology*.

Considering the other end of the spectrum, it should be understood that the concept of *Spirit* and *Divinity* is not excluded from the holistic *Systemological paradigm*.[92] However, in contrast to any "religious" conception of "God" or "Heaven" or a "Divine Being"—all of which we leave to the *Seeker* to determine appropriate definitions for—the idea of a "Spiritual Universe" (AN) and the "Spiritual Infinity" (AB.ZU) is treated to a higher degree of significance than any other formal spiritual science or philosophy. The type of "Causeless Cause" or "Infinite Spiritual Divinity" expressed mathematically—and in the *Standard Model*—as "*Infinity*" (8.0) cannot be reduced to a form or personified or "anthropomorphized" into in any aspect for us to behold in *beta-existence*.

It is only when we conjecture and postulate to define and limit the *"Infinite"* that we develop erroneous dogma and religion—where one individual maintains some inconceivable *belief* as a solid "thing" to use as a basis for other beliefs. Your very being and *Lifeforce* is your power of "faith" and it is YOU that chooses to *cement* it in static[93] beliefs *or* use it to further you up the *Pathway* toward your spiritual *Ascension*.

---

92 **paradigm** : an all-encompassing *standard* by which to view the world and *communicate* Reality; a standard model of reality-systems used by the Mind to filter, organize and interpret experience of Reality.
93 **static** : characterized by a fixed or stationary condition; having no apparent change, movement or fluctuation.

THE PHYSICAL SIDE OF
SYSTEMOLOGY PROCESSING

A greater understanding of the "Physical Universe" (KI) is also achieved with effective *Processing*. It is important that a *Seeker* does not ignore the significance of this *beta-existence* and their participation in its Reality. First and foremost, however, it is important to consider the way in which *any* aspects of the "Physical Universe" *are* handled by an individual, including everything from the manner in which they take care of their "*genetic vehicle*" and all other possessions within their range or sphere of direct influence, such as a "*mechanical vehicle*" (car or truck). This includes the health or condition of physical objects and also the pattern of which they are kept, arranged, organized—or a lack thereof.

An individual is likely to maintain their level of *Awareness* in line with the manner of responsibility and organization taken toward the management of physical affairs. Tendencies, inclinations and behavioral patterns all generally follow a predictable array, depending on the very state of *Awareness* being maintained. A person negligent about management of thoughts, emotions, and other *Lifeforms* will likely be negligent in personal care and of possessions. This is important not because of some strange moral lesson, but as an observation of results concerning levels of *Awareness* maintained during these states of *indifference* toward existence.

—RECALL a moment when you discarded something you later needed. *What is it? Where are you? What do you hear? Who else is there? What emotion do you feel after discarding it?*

—RECALL the moment you realized you needed what was discarded. *Where are you? What do you hear? Smells? What else do you see? Who else is there? What emotion do you feel?*

—RECALL a moment when you were forced to have something. *What is it? Where are you? Who is making you have it? What are they saying? How large is the space? How bright is the space? What emotion do you feel?*

—RECALL a moment you discarded something you were forced to have. *What is it? Where are you? Who made you have it? What time of day is it? What emotion do you feel after discarding it?*

—RECALL a moment when you discarded something you didn't want. *What is it? Where are you? What do you hear? Who else is there? Smells? What emotion do you feel after discarding it?*

—RECALL a time you enjoyed throwing away something old or broken. *Where are you? Who else is there? What do you hear? What time of day is it? What emotion do you feel?*

Actualizing Awareness also requires recognizing when you are not *Aware*—or are unaware. Some will have difficulties admitting or even accepting this. If maintaining full *beta-Awareness*, the *Self* is directing with total control as the *Alpha Spirit* making the best choices and processing information at optimum efficiency for management and operation of *beta-existence*. However, when there are numerous cemented *imprints* and *thought-forms* displacing and filtering our clear *Awareness*, suddenly there is an "unawareness" at play—and some aspect of what is being perceived as Reality is actually *fragmented*.

Fragmentation can only result from *unawareness*—when we are not in control of our "creations" and later we can realize that these "reactive-response" mechanisms that engage are actually quite destructive. When we untangle and resurface these mechanisms with *analytical recall*, they are no longer able to siphon off our attention-energy completely *unaware*. When we speak of fragmented *imprints* and *thought-formed beliefs*, these are not the same as the powerful creative imaginative faculties that an individual innately possesses and which they gain skill in accessing vividly as more of their personal stores of ZU energy are freed up. Anyone with developed critical thinking and imaginative skills should be able to fashion and manipulate and erase any variety of mental images and perform "thought experiments" without agreeing to a *fragmented* Reality as a result and becoming its *effect*.

Those individuals that are "unaware that they are unaware" are most likely "hovering" about existence in their *Lifetime*, courtesy of the "RCC" or *primitive mind*, but otherwise restricted in their ability for *Self-direction*. All those sharing the Human Condition are participants in the creation of this Universe as *it* unfolds; but unfortunately those who are unaware of their participation are just as capable of agreeing to erroneous manifestations and contributing to emotional reactions and thought patterns that do not contribute positively toward their existence or that of others. Most of us know at least one person that demonstrates *fragmentation* of physical "unawareness" in their environment. An increase of *accidental occurrence* just seems to "happen" around these individuals when they "dip" down on the *Emotimeter*—or suddenly new challenges and barriers repeatedly manifest in plain sight that now need to be overcome, when such would otherwise not exist. In extreme cases, the sheer likelihood that some of these incidents happen at all would even border on the *"supernatural."*

Beta-Awareness developed along with other genetic faculties of the *genetic vehicle*. These aspects are related specifically to the Human Condition and not the *Alpha Spirit*, which is not very *fragmented* as a spiritual *Identity*,[94] but of which receives a *fragmented* experience of *beta-existence* as a result of *imprints* and *thought-formed beliefs*. These are not necessarily a *causal* creation of the *Alpha Spirit* carrying it, but *It* agrees to the information and thereby participates in precipitating[95] the Reality and its generation of form and solidity. An *Awareness* of an external Universe taking place "outside" of *Self* is what allows us to "create" from a very basic point of participation. That *Self-directed* participation increases as the certainty of our "knowing"—or state of *"Knowingness"*—increases. As a result, we develop a "sense" about the "outside world" independent of our *Self*. This is what some spiritual philosophers

---

94 **Identity** : the collection of energy and matter—including memory— across the *"Spiritual Continuum"* that we consider as "I" of *Self*.
95 **precipitate** : to actively hasten or quicken into existence.

refer to as "simple consciousness" and it reflects basic primitive functions maintained at (2.0) by the "RCC."

The basic "primal" physical drive to exist is "terrestrial" in its programming, developed from millions and perhaps billions of years of cell division and genetic memory on planet Earth. This "simple consciousness" is linked to the basic "animal mind" governed at (2.0) and is linked to the heart and solar plexus region of the physical body. This part of our "*genetic vehicle*" is linked to, or rather, carries "survival information" all along down the "genetic line"—what it contains is *not* a record of the "past lives" of the *You* that is *Alpha Spirit*, but rather a record of the organism's own genetic survival on the planet for however many millions of years it has been "conscious" and dividing.

Basic *beta-Awareness* of the Physical Universe was once governed by the RCC exclusively before it evolved (or was upgraded) to a higher point that could accept *Alpha Spirits* to contact and control it from a higher level "control center"—the "MCC"—at (4.0). This higher "control center" is not *You* either, but it is the point of contact, or basic "seat of consciousness" for the Human Condition—operating around the head region of the physical body attributed to the "Third Eye." This is a "command center" only—the actual presence of the *Alpha Spirit* is primarily exterior to the borders of the physical body. The *Alpha* presence emanates outward from the body approximately to the degree of the "ethereal aura." We would expect to see an elevation of *this* "seat" for a fully actualized *Homo Novus*. Some have suggested a minimum point of WILL (5.0)—the level of actualization marked by total appreciation of aesthetics, art, beauty, creativeness, design, symmetry, &tc. There are others that believe the next shift for true spiritual evolution is the move "through and out" to *Infinity* (8.0).

IF that is the case—we sure have a lot of work to do.

SELF-PROCESSING ATTENTION PATTERNS

Self-directive ability of the *Alpha Spirit*—the actual "YOU" that is doing the directing of the *beta-experience* whenever your RCC programming isn't "kicking in" and displacing your clear channel of *Awareness*—is of such a high degree and frequency that it could not possibly all be contained within a single physical body within a single lifetime. Nevertheless, the "primitive mind" of the Human Condition (when activated) does not necessarily promote, or even allow for, a "higher" faculty of "consciousness" to be simultaneously maintained. Individuals who maintain only lower levels of *Awareness* are not operating as though they are actually *Aware* that they are "Eternal Spiritual Beings." This behavioral tendency and thought pattern increases the closer an individual's state gets to matching personal frequencies with the Physical Universe on a material level around (1.0) or lower on the *Awareness Scale* or *Emotimeter*.

Below (2.0) in *Awareness* an individual is drawing energy in to their *imprints* and *thought-forms* as opposed to directing and projecting energy outward in a circuit. The further below (2.0) an individual reaches in their emotional processing of existence, the more of an energetic "black hole" they become—not only to themselves, but to their environment—*until the serpent is finally successful in devouring its own tail in entirety*.

"Behavior Modification" techniques and "Motivational Coaching" became a large feature of the alternative, "self-help" and "New Age" movement that rose in popularity during the later half of the 20th Century. Most of it stemmed from a revival of "Eastern philosophy" in the Western world, which also accompanied the underground "New Thought" movement, and other forms of "creative psychology," that became the subject of classrooms and bestseller lists for multiple generations. An individual could even graph a link to support the rise in demonstrations and applications of these subjects as a direct counter-effort to the rapid industrialization of Western civilization that brought us to the doorsteps of an electro-cybernetic digital age. Spiritual and Physical (external) Technologies compliment each other quite profoundly if strengthened and coordinated together holistically. *NexGen Systemology* is arriving at this time to make certain this course does not remain fixed in its "*one-sidedness.*"

—RECALL a moment when someone *did* something that confused you. *Who is it? Where are you? What do you hear? What do they say? What position is your body in?*

—RECALL a moment when someone *said* something that confused you. *Who is it? Where are you? What do you hear? What do they say? What time of day is it?*

—RECALL a moment when you were really focused on something. *What is it? Where were you? What actions are happening around you? Is anyone else there? What position is your body in?*

—RECALL a moment when you focused your attention and were suddenly distracted. *Where are you? What were you focused on? What broke your attention or confused you? What do you hear? What else do you see? Who else is there? What do you smell?*

—RECALL a moment when something suddenly fixed your attention. *Where are you? What focused or fixed your attention? What actions are happening around you? Is anyone else there? What do you hear? What else do you see? What do you smell?*

—REPEAT the *last two* "command lines" several times, alternating between, then end the session.

Orientation of *Self* as present in *time* and *space* is an important feature of *Awareness*. Perhaps at its most basic physical level of association, an *Awareness* of "present time" is primarily what most people associate with "attention." For what else is personal *Awareness* when applied to the management of *Self* and *environment*, but *attention*? The idea of "consciousness" and *Awareness* is often a vague fluffy idea for some individuals to grasp—but surely you can then better understand the concepts presented in this present volume as a selective *Self-direction* of personal "attention."⁹⁶ There is a popular *NexGen* axiom* we started to use when *Systemology* began that states:

Energy flows where attention goes.

When we bring our *Life-experience* up to *analytical levels*, we find that our accumulation of experience takes on a much "lighter" presence in our field of vision than when we choose to make it a part of our "baggage." If you want to take that on along with you as part of your journey, the way forward will often become too cumbersome—too monotonous, strenuous and/or antagonistic—for the *true realizations* to be actualized along the way. It is for this reason that many *Seekers* halt their own progress: there are too many fancy lights and buzzers to otherwise occupy our *attention* that we find ourselves—as we increase in age and accumulate untreated "experience"—more and more the *effect* of someone else's *cause*. As long as our *Will* does not succumb to such drastic levels, we have no reason not to maintain vital health and longevity in this lifetime. As soon as we begin to believe that its *futile* and its *all over*, well then it is—or it will be soon—and we can easily see many of these people simply waiting for this *end* as they go about in the world. It is directly demonstrable in the behaviors and attitudes that they reflect outwardly in the environment.

Directing attention is one of the most basic applications of *Self-determinism*. However, in this instance, by "basic" we mean both: "simple to master" and "fundamentally essential" to personal progress. An individual comes into the world full of this ability to direct attention because everything is "new." Things carry, what is called, the "novelty effect" of being interesting or occupying our attention solely for the fact that

---

96 **attention** : active use of *Awareness* toward a specific aspect or thing.
* Presented by Joshua Free at the 2009 *"Reality Engineering"* lectures.

they are "new." As we go through a normal lifetime or displacing our *Awareness* of the environment with the *imprints* and *thought-forms* that we *believe* our environment to be composed of, we increasingly lose our directed attention as *Self* and therefore give up the "ownership" and "responsibility" of determining *Self*, Reality and the Physical Universe (KI) around us. This is sometimes demonstrated to the extreme with an individual "constantly needing to buy something new" and disregarding former possessions. The attention, energy and resources—not to mention "buying power"—for a single lifetime becomes distracted exclusively on "mastering" various levels of *Havingness*, and so long as that is where they remain in *Awareness*, the "black hole" or "void" is unable to be "filled" or satisfied.

One of the techniques used in sessions by *Pilots* to align the *Seeker's* attention for *Systemology Processing* is referred to by different names or "commands" from different sources and preferences, but generally relates to "spotting and identifying objects in space." This is often used prior to deep meditations and spiritual rituals in various traditions, whenever a practitioner is seeking to "ground and center" themselves—their *Self*—in the present moment with full *Awareness* and attention. It should seem strange that we should get by very well without this state—and there are many Humans that obviously do not. Our state of *Awareness* and attention is directly related to the "brightness" in our lives. As we become too familiar and desensitized to the objects and spaces around us, our *Awareness* "dims" and so does our perception and management of Reality in the Physical Universe.

During the original "RR-SP-1" extended course lecture—of which a transcript appears in the text: *"Tablets of Destiny"*—the present author explains:

> *Processing* begins with entering a state of focused concentration, which for most people experienced with

meditation and thought discipline from former Grades, is as simple as a few deep breaths and closing their eyes. However, it may be that the *Seeker's* current *Awareness* level maintained on the ZU-line is simply providing too much restlessness or withdrawal to even begin a process designed to raise levels of *Awareness*. This happens all of the time and may be remedied without breaking the session. A common exercise to refocus attention before starting a "process" is for the *Pilot* to have a *Seeker* look around, spot and identify basic objects in the room. This helps orient the *Seeker* to be "in phase" with *Self*—which they should always be at the beginning and end of a session—and preferably also during in an ideal Reality.

The resolution for this *attention* on *Havingness* is processed very simply with "spot-identify" alternations of *Awareness* in the space we occupy. This seems rather silly and childish to some, but a simple test will show its effectiveness. In fact, the effectiveness is so sharp that this may be used at the beginning *and* end of a session if desired, or also at any point during *Self-Processing* where you feel you may be digging into something or some aspect deeper than you are ready to handle. There is no actual danger to any *Systemology Processing*, because even in the most turbulent times a *Seeker* merely returns to the points of *Awareness* they are most certain of—in order to stay "in phase"—and this can include exercises as simple as seeing, identifying and touching something "you can have." It is not enough to just glance—really focus your attention on the object, notice its points of solidity and identify it at an analytical level. Then touch it and analyze any additional *facets*.

—LOOK around you and SPOT an *object* in space. IDENTIFY its *solidity* and CONTACT its *substance*.

(*Do this several times with various objects, walls and corners, until completely "in phase" and interested or extroverted about your Path.*)

This is "selective attention" pure and simple—and next to the regulation of *breath*, personal selection of *attention* is one of the most critical faculties for *Self* to *determine*. When you first begin to practice this in sessions of *Self-Processing*, you may use simply the first part. However, the second step of advancing this technique is virtually the same "command"—but there is a particular difference in the "technique application."

—LOOK around you and SPOT an *object* in space. IDENTIFY its *solidity* and CONTACT its *substance*.

(*Do this with the same objects, walls, corners, &tc.*—EXCEPT remain seated with eyes closed.)

DEFRAGMENTING "SELF-CONSCIOUSNESS"

In previous chapter-lessons within this volume and former texts, we have demonstrated and previewed the nature of "basic genetic consciousness" or the "primitive mind" that carries a *store* of our *emotionally encoded imprints* as memory-data for "reactive-response" programming. *Defragmenting* the data-banks at the lower levels operated by the "RCC" raises personal *Awareness* in critically significant ways with otherwise unparalleled results. It is also important that we directly address "higher" faculties governing "higher" frequencies on the ZU-line of *"analytical thought"*—as governed by the "MCC" *Master Control Center* for *beta-Awareness* at position (4.0) on our models.

If we are treating "reactive degrees" of sub-par *beta-Awareness* on the ZU-line between (0.05) and (2.0), then the range of thought activity between (2.1) and (4.0) would be "active degrees" of *beta-Awareness*. It is only within this band—from (2.1) to (4.0)—that we actually maintain any degree of true "Self-Consciousness." In fact, it is this very division point that conventional sciences have debated about concerning the "consciousness" or *Awareness* potential of animals and plants.

We can essentially demonstrate that all *Life*—all expressions of *ZU* in the Physical Universe—maintains at the very least a "primal/primitive mind" of which only environment and evolution may develop. Based on our current standards of gradation, we can assume that animals actually maintain a higher range of *Awareness* than Humans often give up credit for. If for no other reason, this *Awareness* in animals would even be able to be advanced further with "right experience" in proximity to *Self-Honest* Humans. This is only theoretical at present and remains open for future *Systemology* work to evaluate on any scales.

When an individual maintains *Awareness* within the band of "active thought"—between (2.1) and (4.0)—there is some recognition of actualized *Awareness* extending "above" the lower levels of purely physical matter and emotional reactions. Even if greatly *fragmented*, within the range of beta "Mind-Systems," all energetic activity reflected by the *Identity* in beta-*Awareness* from *Self* is to the tune of: "between 25% and 100% consciously *Self-Determined*."

Unless the "misemotional" states are only visited temporarily as a part of a session, any such lesser level of *Awareness* than "(2.0)/25%" renders *Analytical Processing* effects negligible.[97] There must be some significant *Analytical Awareness* in play to process information, perception and experiential memory "*analytically*." And everything short of this is hardly even "valid memory"[98] because it is being stored as "erroneous programming" and "conditioning"—much of which has been enforced on us with efforts literally *against* our WILL. Literally. This is what we are *Processing-out*—"resolving"—to regain an actualized state representative of the original, basic power and ability of WILL. In fact, you may wish to engage *Analytical Processing* on this now.*

---

97 **negligible** : so small or trifle that it may be disregarded.
98 **validation** : the reinforcement of agreements of Reality.
* For improved effectiveness and so the *Seeker* may apply a wider range of work to their individual case, suggestions for recall of specific *facets* will no longer be added to the *Processing Command Lines*. A

—RECALL an incident when you were invalidated by someone.

—RECALL an incident when you invalidated someone else.

—RECALL an incident when someone invalidated someone else.

—RECALL a moment when you were suddenly interrupted by someone.

—RECALL a moment when you interrupted someone else.

—RECALL a moment when someone interrupted someone else.

—RECALL the last time you were told to know something important.

—RECALL the last time you told something important to someone else.

—RECALL the earliest time you were told a secret.

—RECALL the earliest time you told a secret to someone else.

—RECALL an incident when a belief was enforced on you.

—RECALL an incident when you enforced a belief on someone else (or others).

—RECALL an incident when someone enforced a belief on someone else (or others).

—RECALL the earliest time you told a secret to someone else.

—RECALL a time that someone lied about you.

—RECALL a time that you lied about someone else.

—RECALL the last time when someone lied to you about someone else.

—RECALL the last time when you watched someone lie to others.

Seeker should take time to review and practice from the *facets list* provided earlier—apply any and all relevant perceptions to *Processing* as applicable.

—RECALL a time that you lied about someone else.

—RECALL a moment when you were picked or chosen.

—RECALL an event you won at or a time when you were best at something.

—RECALL the last time when you discovered something to be untrue.

—RECALL the last time when you discovered something to be true.

—RECALL a time when you were right after all.

There are many instances when individuals will decide—or agree to—Reality inaccurately outside of *Self-Honesty* as a simple result of programming. In most cases, it will always seem to us that we are fully in control when making our decisions, but too often the information used as a determinate is itself inaccurate or erroneous. Most *fragmentation* experienced in *beta-Awareness* is not a result of the *Alpha* state of *Self*, but from others that we share experiences with in *beta-existence*. While it is true there are higher "spiritual" levels of *fragmentation*, it should suffice for the majority of those journeying on the *Pathway to Self-Honesty* to first treat specifically the "physical/emotional" and "psychological/mental" degrees of *beta-fragmentation*. A *Seeker* is directed to reach a conscious realization of their management in these areas first —especially as a tenant of the Human Condition, since these *Games* are encountered most frequently on a daily basis.

Of the many ways in which *beta-existence* may diminish willingness (and capability) to *Self-direct* with certainty, the most critical *fragmentation* consistently received from others—and social environment—may be reduced to two main categories:

a.) *Invalidation*;[99] and

b.) *Enforcement*.[100]

---

99 **invalidate** : decrease the level or degree or *agreement* as Reality.
100 **enforcement** : the act of compelling or putting (effort) into force; to compel or impose obedience by force; to impress strongly with

Both methods involve strong emotionally charged communication of *intention*, *effort* and *belief*—and both "wave-forms"[101] of energy operate outside of *Self-Honesty* as a "fallacy"[102] purely for the fact that they come from (allegedly) "authoritarian" sources. Any "appeal to authority" in logic is a "fallacy" and therefore *fragmentation* by definition.

A *Seeker* that has progressed to higher vibrations of *Awareness* is better equipped to manage the *enforcement* of "knowledge" and the *invalidation* of "belief." Yet, along the way to this actualization, *Self-Processing* is used to release the *Seeker's* stores of emotional energy tied to *former* experiences of *invalidation* as a means of increasing ZU-frequencies applied to present and future management of the same types of situation. When someone is approaching an individual from the point of enforced authority, they are installing or *imprinting* their own emotionally charged *beliefs* by reducing the perceived *Awareness* level of the individual. The basis for this reduction may be as *material* as the threat of violence or it may be socially learned and conditioned to simply "obey authority."

Invalidation—and the information it carries—by definition, is a low-energy vibration or emotional solid *imprint* that is presented as fact. *Fragmentation* occurs not only when this happens *to* us, but also when we have invalidated *others*. In a state of objective *Self-Honesty*, such as what is demonstrated by a *Pilot* in *Systemology Processing*, an *Alpha* should never "invalidate" *or* "validate" the experiences of another.

---

applications of stress to demand agreement or validation.
101 **wave-form** or **thought-wave** : a proactive *Self-directed action* or reactive-response *action* of *consciousness*; the *process* of *thinking* as demonstrated in *wave-form*; the *activity* of *Awareness* within the range of *thought vibrations/frequencies* on the existential *Life-continuum* or *ZU-line*.
102 **fallacy** : a deceptive, misleading, erroneous and/or false beliefs; unsound logic; persuasions, invalidation or enforcement of Reality agreements based on authority, sympathy, bandwagon/mob mentality, vanity, ambiguity, suppression of information, and/or presentation of false dichotomies.

Our goal is not to constantly impress our viewpoints on others, but also we should maintain the high-frequency states that better galvanize our core as *Self* from the invalidation and authoritarian enforcement of others. It is also important that we only "evaluate" (not necessarily "validate") our own certainty for our *Self* and then "encourage" others on their journey toward *Self-Honesty*—but this is accomplished best without cementing more premises and axioms for Reality than are strictly necessary to keep the universe intact.

THE GAME OF "SELF-CONSCIOUSNESS"

The paramount signature *Game* of the actualized *Alpha Spirit* is an ability to create and un-create at *Will*. At every turn in *Systemology* we find increasing realization of abilities and education regarding applications of *Games* and *Systems* to the management of our *beta-existence* and our environment—and every step of the way we are working successively[103] toward the *Actualization* of the *Alpha Spirit* as "*I.*" All of these aspects, conditions—and even the very *Processes* themselves—point toward one key theme:

CONTROL

It should not be altogether surprising that a societal civilization so *fragmented* on the concept of "control" should produce populations of the Human Condition that maintain exceptionally low levels—or *no* level—of "*Self-Control.*" The very idea of "control" is *necessary* to existence—the very operation by which all that we might *do, think* or *be* is directed from *Self*. As with many other necessities of *Self-Honesty*, the "reactive-responses" and inhibitions concerning "control" are most likely the result of improper management of "control" as conditioned experience: *toward* you and *by* you.

Our understanding and energy reserves regarding "control" are so deeply fragmented that the Human Condition has been

---

103 **successively** : what comes after; forward into the future.

left with very little of its own *Self-control*—meanwhile it is being so thoroughly dictated and enforced from outside sources. Even the very idea conjured to mind from the word "control" is likely to spark some emotional reactions and beliefs—which will have to be *processed "out,"* particularly if it is too difficult to "associatively" or "analytically" change by personal command—which is yet again, another demonstration of *Self-directed* control of *Self*. Technically, "AR-SP-2" methods may be used to *"Process-out"* associated emotional charges and personal ZU stores embedded with any "word," "concept" or "idea" that is reinforced by *Self* or programmed as a "belief." For example, the very idea and phrase "self-conscious" even seems to register negatively for many individuals if following standard issue social programming. The same is true for many *"keys"* to our success on the *Pathway* that are hidden in plain sight.

Validation and Invalidation of Reality is subject to the degree of *agreement* between individuals far more than any particular *facts* involved. The "facts" are, in effect, whatever we *agree to*. The certainty of our agreements and their demonstration in the Physical Universe (KI) is what generates our level of knowledge and understanding, which is also related to our *Awareness* level. All of these aspects work together systematically—and as a "continuous whole"—which is why Reality seems as *solid* as it does. Especially at the most physically material levels, there is less degree of *motion* and *free movement* available. This is why the substance of the "Physical Reality" appears so fixed at the levels we interact with in "touch" &tc. around (1.0) and lower, when we are directly starting to contact the material of the (*Beta*) Physical Universe (KI) with the material of the *beta-Body*. This is also why *enforcement* of Reality communications at this level are *so* "physical"—whether through the violent acts of high-frequency emotion or the soothing coos of low-frequency emotion, the *enforcement* is simply a *"re-inforcement"* from an outside source as opposed to our own *determinism*.

Invalidation is generally different from *enforcement* because there is some energetic connection between you and the source—such as the recognition of authority or the respect for a mentor or loved one. The communication of "invalidation" is not received directly as *fear* or *anger*, which would be more along the lines of *enforcement*. Rather, it is an appeal to our intellectual faculties within the range of *Thought*, which we then "process" and reinforce with our emotional reactive energies from within. This is how a "belief" is crystallized or cemented. So long as *thought-forms* are kept under *analytical control* of *Self*—to be created and un-created at *will*—there is no concern of heavy fragmentation, because the individual is operating at such a high-frequency of ZU that the energy of these forms may be just as easily dispersed, freed and reassigned. That is the type of *fluidity* that we seek to actualize as *Self-Honest Thought* because it promotes the highest creative potential of *Self* throughout all dynamics of existence.

In the range of thought—under control of the Alpha "MCC"—ZU energy is easily projected and expressed in *beta-existence* as managed by the *Alpha Spirit*.

Within the range and control of the "RCC"—or *Primal Mind*—personal ZU energy is turned inward and collected as *fragmented beliefs* and *imprinting* below the intellectual surface.

From a point of *Self-Honesty* we can actually use the Mind-Systems to effectively process knowledge and evaluate our certainty to manage Reality. This leads to the ability to perform—what Albert Einstein referred to as—

> Thought-Experiments : using faculties of the Mind's Eye to *Imagine* things accurately with considerations that *have not* already been consciously experienced by the body in this lifetime (*beta-existence*).*

---

* "Thought-Experiments" (from the German, *Gedankenexperiment*) are logical considerations or mental models used to concisely visualize consequences (cause-effect sequences) within the context of an imaginary or hypothetical scenario. Einstein's most popular theories

True creativity, innovation and development is not a result of duplicating existing *solids* and cemented *beliefs*—although contemporary educational systems operate as though they do. It is necessarily the application of "Imagination"—which is one of the highest frequencies of ZU energy usually tapped by the Human Condition during a lifetime, and it does not even fall within the band of *Beta*. This is one way in which we may perceive with some certainty that there is "more than" the Physical Universe as we experience it "physically."

—IMAGINE a community of humans openly *validating* beliefs of every individual 100% of the time.
ANALYZE+EVALUATE its potential for success, creativity and innovation.

—IMAGINE a community of humans openly *invalidating* beliefs of every individual 100% of the time.
ANALYZE+EVALUATE its potential for success, creativity and innovation.

—IMAGINE a community of humans required to *validate* the beliefs of one individual 100% of the time.
ANALYZE+EVALUATE its potential for success, creativity and innovation.

—RECALL the last time when you were stopped from finishing something.

—RECALL an instance when you stopped someone from finishing something.

—RECALL a moment when someone stopped someone else from finishing something.

—RECALL a time that you successfully completed a task.

—RECALL a time when you were satisfied with the results.

When an individual expresses free use of "Imagination" they are actually tapping power inherent in the *Spirit* and not the *Body*. It is this unique and advanced *Alpha Spirit* that has actually allowed the Human Condition to survive and progress at

---

resulted from thought-experiments involving "riding beams of light."

is has—and this range of frequency operates above (4.0), around the band of WILL (5.0) on the currently unreleased extended "*Total Awareness Chart.*" As a *Seeker* is moving along the *Pathway* and progressing in their *defragmentation*, more and more channels to higher faculties begin to open up more readily. The reason we first focus specifically on frequencies within *beta-existence* is because receptivity to accessible higher levels of creativity and reasoning naturally increases during this present *beta-Processing* stage. When a person is freed up from their entangled *creations* and *enforced* beliefs, the way forward hastens with efficiency.

MANAGING CONSCIOUS AWARENESS

In addition to basic management of *Self* and Reality, selective application of *attention* as *Awareness* is one of the primary keys to all effective "esoteric" spiritual or mystical work—on any level. It is the "activity" of our *attentions* that people are often describing with the word "consciousness"—which is an otherwise very vague and ambiguous term. In the vocabulary of *NexGen Systemology*:

> <u>Consciousness</u> is specifically defined as—"the energetic flow of *Awareness*; the *Beingness* component of our existence in *Spirit*; the Principle System of *Awareness* as *Spirit* that directs action in the Mind-System."

Our ability to direct *attention* and apply personal energy toward existence is related to our gauged levels of *Awareness* and certainty on the *Pathway*. Although we each begin a physical lifetime with vast unlimited personal potentials, this diminishes over time only as a result of *creations* that we either have had impressed upon us, or otherwise have developed ourselves, but have simply forgotten. Our *creations* do not just "go away" due to abandon or neglect. Our resources and personal stores went into *creating* them—and until that energy is recalled and reclaimed, it remains tightly wound to form the *solids* of our belief imprinting. Even turbulent energ-

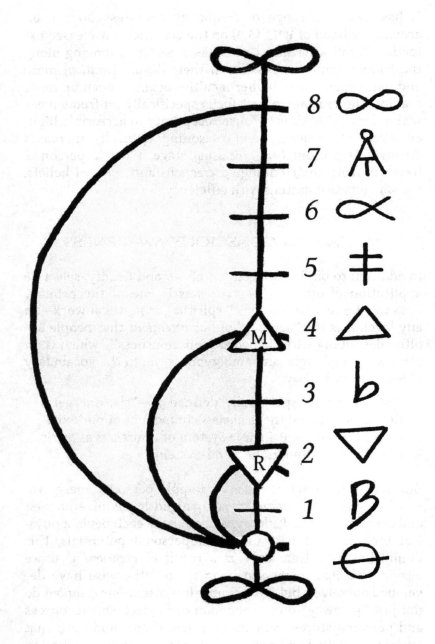

etic encounters with others do not successfully "destroy" our creations—often they simply add more and more layers to them, adding lines of code—counter-thought programming—to an existing *solid*. This simply makes what we carry with us *that much* heavier to behold—and as a result, we get into a behavioral pattern of not being *expressive* and withholding all of our collected energies until we buckle beneath their weight. The unburdening of this weight is the sense of "release" or "relief" experienced during *Self-Processing*. We are, in all actuality, literally "sighing" with *relief* as we discharge stores of inhibited personal ZU energy and raise our *level* of *consciousness*.

We can register a certainty of our description of ZU for all levels of "Self-Consciousness" using information from *NexGen Systemology* knowledge lectures and the *Standard Model* of *beta-Awareness*:—

••• The simple range of "physical consciousness" maintained by the "RCC" (*Reactive Control Center*) of the "primal mind"—which includes emotion and reactive programming—describes all personal ZU activity of the *genetic vehicle* at (2.0) or below.

••• The analytical thought range of "mental consciousness" is maintained by the "MCC" (*Master Control Center*) —which includes an allegorical "facsimile" of *Self* as a "*beta-personality*" developed from the myriad of *imprints* and *programming* carried by the Human Condition (*homo sapien*) between (2.1) and (4.0).

••• The actualized *Alpha* range of "spiritual consciousness"—or actualized *Self-Consciousness*—is only experienced as *Self* directly "*in Spirit*" from (4.1) and higher, which is *exterior* to the Physical Universe; and we may graphically plot that the *Awareness* level of *Total Self-Actualization* as an "*alpha-individual*" is maintained by an

"ACC" (*Alpha Control Center*)[104] at (7.0).*

••• There is also the matter of non-individuated degrees of ZU as "cosmic consciousness"—extending between (7.1) and (7.9)—that encompasses[105] all substantiated universes and *Lifeforms*, separating the ALL from the *Infinity of Nothingness* at (8.0) on the *Standard Model*.

The present Grade of *Systemology* instruction and *Self-Processing* emphasizes *defragmenting* the entire range of *beta-Awareness* up to (4.0) as a benchmark for basic *Self-Honesty*. This would effectively clear out heavy *fragmentation* of any personal energies along the ZU-line pertaining to "physical" and "mental" consciousness. A primary element of this consciousness actualization is the ability to maintain *Awareness* of *Self* and "orient" it as "*present knowingness*." This is written in older *Mardukite Systemology* volumes as the "*kNow*"—derived from the phrase "in the know"—but pronounced as "know" or "now" by personal inclination since they are both referring to a "present state" of *actualized* "*knowingness*."

Although it may seem trite to some readers, the alignment or orientation of *Awareness* "in phase" with present *Self* or "I" is not often the level of *actualized Awareness* maintained by the Human Condition. Our internal research developed from

---

104 **Alpha Control Center (ACC)** : the highest relay point of *Beingness* for an individuated *Alpha-Spirit, Self* or "I-AM"; in *NexGen Systemology*—a point of spiritual separation of ZU at (7.0) from the *Infinity of Nothingness* (8.0); the truest actualization of *Identity*; the highest *Self-directed* relay of *Alpha-Self* as an *Identity-Continuum*, operating in an *alpha-existence* (or "Spiritual Universe"–AN) to *determine* "Alpha Thought" (6.0) and WILL-*Intention* (5.0) *exterior* to the "Physical Universe"–(KI); the "wave-peak" of "I" emerging as individuated consciousness from *Infinity*.

* Information regarding "*spiritual consciousness*," "*Alpha states*" above (4.0), and specifically the "ACC" (*Alpha Control Center*) is formally introduced to *NexGen Systemology* in this present Self-Help book as based on the original "*Crystal Clear*" series of lectures delivered by Joshua Free to the *NexGen Systemology Society (NSS)* for the *International School of Systeology (ISS)* in October 2019.

105 **encompassing** : to form a circle around, surround or envelop around.

early use of our "BAT" tests and *"Emotimeter"* established some very important conclusions concerning our orientation of *Self* "in phase" with our *beta-experience:—*

- Those scoring *(1.4) or lower* had recently experienced significant "loss"—or they had not successfully recovered *Awareness* and personal stores of ZU from a previous—even potentially distant (time)—experience of "significant loss" or "trauma."[106]

The nature of this *loss* was also determined or found to be connected to *any* perceived "great loss" even if not a tangible item or departure/death of a person or *Lifeform*. Even these incidents are drawing from deeply embedded emotional programming that *fragments* thought as an automatic response to the confusion of information received—which is not *Self-directed*. Likewise, we are programmed socially to "blanket our *Awareness* in grief" in order to be considered "normal"— which is acceptable for the moment of receipt, but not for a person to "give up on their own life" because they could not grant another person the *beingness* of physical *immortality*. In fact the deepest archetypal *imprint* of "Loss" is likely not a "thing" but a "sense" of "diminishing existence that is not *Self-directed*." The event-scene *imprinting* this is usually unique to each individual.

- Those scoring *(1.5) up to (2.2)* had recently experienced significant "invalidation" or "enforcement of Reality"— or they had not successfully recovered *Awareness* and personal stores of ZU from a previous—even potentially distant (time)—experience of significant "invalidation" or "authoritarianism."[107]

---

106 **traumatic encoding** : information received when the sensory faculties of an organism are "shocked" into learning it as an "emotionally" encoded *"Imprint."*
107 **authoritarianism** : knowledge as truth, boundaries and freedoms dictated to an individual by a perceived, regulated or enforced "authority."

We have touched upon invalidation and enforcement in the current lesson-chapter.

> • Those scoring *(2.2) or lower* maintained significant *attention* and *Awareness* on *past events*—even potentially distant (time) experiences—which tend to reinforce existing *imprinting* and *conditioning* without the "present-presence" of *Self.*

The individual is only marginally "*Aware* that they are not *Aware.*" Hence the cut-off point of "physical consciousness" on the *Standard Model* at (2.0)—when the RCC engages directly because the *living genetic organism* is, at that degree, only maintaining 25% *beta-Awareness* from *Self*. As the ZU within a *genetic vehicle* is less *Self-determined*, the mechanical motor functions of the *vehicle* simply begin to operate on base-line "primitive/primal" programming. This is wholly reactive—based on reactive-response programming (*conditioning*) that is both installed genetically and then accumulated and reinforced over the course of a single lifetime—until apparently it is unable to sustain its own existence further while operating on basic programming.

> • Those scoring *(2.3) to (2.9)* maintained significant *attention* and *Awareness* on *present events*—even to varying degrees that they are "stuck" in loops of present problem concerns—but they demonstrate engagement of "mental consciousness" to meet demands currently present in life after first successfully overcoming past challenges.

Although strength (degree) of "mental consciousness" might be assumed to be a result of the development *of* "mental consciousness" faculties directly—it is actually by freeing up the energy maintained *below the surface* of "analytical thought" that will raise the level of *Awareness* or "consciousness activity" up into the range/level (band) of "Thought." The gravity inherent in the mass accumulated at lower levels is what "pulls" our *Awareness* down like a solid. Otherwise—in its natural state—*Awareness* is much more efficiently regulated from "higher" levels of control. Therefore, it is a series of actions taking place "against *Self-determinism*" that consistently lowers our level of *Awareness* by keeping it fixed in place with the attentions and emotion given in the past. The methods given as *Self-processing* are utilized for this specific purpose.

Many of the "thought-loops"* are the result of *uncertainty* regarding personal management of environment and sustenance of one's own responsibilities toward continuing a material existence. As we go up and down the slope or our emotional tides, the RCC faculties—or rather their intensity—continuously cycles periods of "engaging" and then "cutting out" rapidly, leaving us with a badly *fragmented* perception of the present and our ability to direct the future. Most individuals that accept they have "but one life to live" will seldom actualize a much higher state of *beingness* in this lifetime—which is a form of irony in itself, for it negates the ability to actualize the *Alpha Spirit* in *this* lifetime. When fully actualized during this lifetime, the *Alpha Spirit* consciously *Self-determines* the "I" into whatever existence it so chooses—by *creating* its next form—and even the universe it exists in. There is no reason this should not be *realized* during this lifetime, because it is in this *beta-existence* that we relearn how we somehow managed to do—*or be subject to*—this exact same thing *here now* in the Physical Universe.

> • Those scoring *(3.1) to (3.6)* applied significantly more *attention* and *Awareness* toward *future events*—increased critical skills of "mental consciousness"—once first actualizing a certainty to meet present demands in life and past challenges.
>
> • Those scoring *(3.7) or above* demonstrated intensive enthusiasm for *Life* with nearly all of their *attention* and *Awareness* applied to *future events*—including a strong interest toward innovation, potential and evolution in the

---

* Referred to as *"figure-8 loops"* in *Liber-One*.

future; development and expansion of *Systemology* for the well-being of Humans; and a genuinely healthy respect for all *Lifeforms* on the planet.

It is in this upper range of "mental consciousness" that we are most interested in actualizing with the *Grade III methodology*. It is an entirely realistic goal within the grasp of all those individuals and *Seekers* that have adopted this "Human Condition" as their *avatar* for *beta-existence*. It is the present authors goal of this current Grade to encourage *Seekers* to "blow out the top" of the *beta-Scale* and assist in ushering in the Crystal Dawn for a new evolution of Humanity as *Homo Novus*—a "new human condition" marked by *Self-Honesty*.

SELF-AWARENESS OF SELF-CONSCIOUSNESS [EXTENDED COURSE]

This is a transcript of a lecture given by Joshua Free on the evening of October 25, 2019. Content of this lecture introduces course material retained by the NexGen Systemological Society (NSS) from the International School of Systemology (ISS). It is included within this volume to supplement previous chapter-lessons and for the benefit of Seekers (and Pilots) interested in experimenting with practical applications of Mardukite Zuism and Systemology (ZU)-Tech alluded to throughout this volume. Additional textbooks, guides and manuals are currently in development to advance further work within this applied spiritual philosophy of NexGen Systemology.

∆ ∆ ∆ ∆ ∆ ∆

"The first step of establishing the *Self* is acquiring a clear, distinct, positive and absolute realization that the *Self* is not the body or physical organism, but is superior to and master of them. Even those students who have entered the plane of 'Mental Consciousness' require additional drilling in order to escape completely the bonds of the physical body."
—William W. Atkinson, a leading pioneer of
American "New Thought," in his book
"Secret Formulas of Mental Alchemy"

Tonight, I would like to share with you some critical information you will need to know in order to understand what is taking place when we apply *Systemology Processing* to defragment "mental consciousness." Most of us by now are already quite familiar with the low-energy ZU-states. We've all experienced these emotions. We've all "been there" so to speak. And that's fine—so long as we don't remain fixed there. What it actually does is allow us a better sense to recognize these states in ourselves and others while *processing*. It is one of the many reason we treat the subject of "emotional encoding" and *imprints* before demonstrating any higher frequency "mental faculties" because these lower emotional states—and the energies and *imprinting* attached to them—are far easier or more accessible for a *Seeker* to recognize, contact and play with; they are much more *physical*, tangible "solids" to begin with. As a result, they also seem the most *volatile* regarding observable physical expression—or dare we say: "behavior."

We speak a lot about the state of *Self-Honesty* in *Mardukite Systemology*—in fact, I believe my first use of the term is in one of the introductions to *The Complete Anunnaki Bible* released not long after the inception of *Mardukite Zuism* and the first active group of *Mardukite Chamberlains* in 2009. With it, my intended purpose was to imply a *perfected state of Knowingness*—which is actually our most basic state of *Knowingness* when we consider the perspective from *Self* as *Alpha Spirit*. Obviously we have a tendency to understand and graphically display things logically from the "ground-on-up." We establish several logical premises before we draw a conclusion—and one of those conclusions is the fact that there is a sequence and order to continuity, however chaotic it may seem when we are placing *Awareness* at the epicenter[108] of *motions*. So, the *Standard Model* is drawn objectively using our normal physical orientation of space to denote "higher" or "lower" aspects between *Infinity* and *Infinity*—or specifically what we show between "KI-*zero*" and "AN-*Infinity*" if looking specifically at the "positive" end of our charted spectrum on the *Standard Model*.

[*Audience Voice:* "*Is there a negative part of the spectrum?*"]

There *is* a "negative" or "sub-ZU"[109] part of the spectrum, yes, which is how we get from *here* ["zero"] on the *Standard Model* down back to an "Infinity" out the bottom. But it is not evident directly in *beta-existence* and would appear to be more along the order of negative space if it were expressed in the

---

108 **epicenter** : the point from which shock-waves travel.
109 **sub-zones** : at ranges "below" which we are representing or which is readily observable for current purposes.

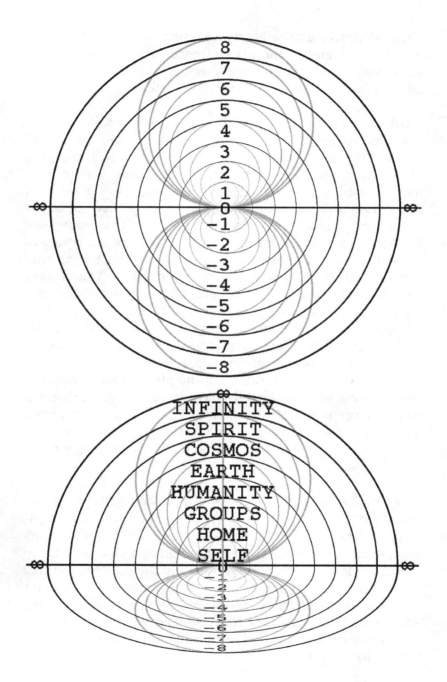

Physical Universe. Therefore, we won't deal with that part directly at present—but yes, we necessarily consider those states when we begin to focus specifically on the *Alpha Spirit* and the Spiritual Universe. Does that...? Okay.

What we are concerned with on the *Pathway to Self-Honesty* is the return to a state that we might speak of in the "New Age" as the "Freedoms of the Spirit." The only issue with speaking of that, or mentioning "freedom" in this sense, is that in its natural state, the *Spirit* is *already* "free." So, to speak of it, as many have done, is to indicate that we are in need of something—or to achieve a state—when we actually already have it. Okay, so it is true that this is a state that we are trying to "actualize" but we are unfolding something that is already present with us. It's something of a puzzle box. You already have it, so you just need to focus and figure it out. People are out there trying to sell you a million distractions away from...
—to just sitting there and figuring it out—and they will describe these distractions or present them or sell you on an *idea* that the inclusion of one more element will somehow make this process of working out the puzzle even easier. Because they know you already have it. And they don't want you to figure it out. Makes them look bad. So they say, here, have one of these pills and you'll just—well you see, you just *got to be* this such and such a person. And we develop what is called: a "personality."

"Personality" is kind of a crappy term to apply to the *Self*—or the *Alpha Spirit* as *Self*—but it seems to apply very appropriately to *beta-existence*. All of society as a whole *really* wants you to *have* a "personality." That should be a clue right there that there is something wrong with the thing. It's funny too, because they just hand them out—they don't even charge you for these things up front. It's one of those amazing gimmicks you see with: "Try this out for a while; but once you assimilate it and forget you have it, well, you know you owe us some monies here, right?"

Now we find that growing up in this society means joining some ludicrous "record club"—*do you remember those?*—a "CD club" or what have you. Everything's digital now I guess. But anyways, they would offer an amazing deal on the bet that you will just start to accept products you would not otherwise buy. And it was quite successful for a long time. I think "Disney" still does something like this now. Anyways—the idea of "personality" is no different. It's about as valuable in the real scheme of things as paper money or any fiat currency and is often what we are earning the money to support—not *Self*, because the *Self* doesn't actually want to be pulled down by these things.

"Personalities" are effects—they are the "ends" and not a "means." They fall under the category of "*things to have*" and therefore must be solid creations of more concentrated energy than *Knowing* or *Being*. "Freedom" is a state of *Being* that is often artificially sold as a "thing" in *beta*—usually by the same authority that is contributing to our *beta*-enslavement or reduction of *Self*. And isn't that just quite the "mafia-game" right there? But "Freedom" is a state of *being*, it is experienced as *Spirit* and it is not something in the category of "*having*" on a material level. No one can *give* it to you. No one can really do anything about *your* spiritual "Freedom" down here on Earth other than *you*. Now, "personality" is a different cat. That's something you can *have* and that's something someone can *give* you—no question about that. It's very solid and once assimilated fully, you would even give up your existence for your right to *keep* it—giving up *beta-existence* for something with no actual form or value outside of the Physical Universe. *Master Control* ain't in charge no more—because that is just not logical. You see: this ZU energy is an amazing thing and once you get a better understanding for how it operates in relation to Cosmic Law it is amazing how many of these natural little ironies seem to maintain its balanced equation as a continuum. You don't even have to get into any maths here to see the effects demonstrated—but it is balanced from a higher spiritual level using "sub-ZU" bands we just mentioned briefly before, which is all tied specifically to our continued *Identity* as *Alpha Spirit* and its ability to create and control a *beta-body*. That's a bit advanced but on the same track as our present discussion on "personalities" and "images" encountered during lifetimes in the Physical Universe.

[*Audience Voice: "Are you saying we should 'Resurface and Reduce' facets of 'personality'?"*]

The short answer to this is: "Yes." But I should note—see, if that scares you—that any concern you have here is purely a semantic one of misappropriated[110] associations. Also, we are concerned with elements of "personality" that are not being *Self-directed*. I mean: the *Alpha Spirit*—you can conjure up a "personality" by your own choosing, if you want, so long as you remain in control of it. That can be part of the *Game* from *Alpha* most assuredly. But, the "personality" is not the "Identity" of *Self*. And the body and personality used for *beta-existence* was likely created by someone else. I mean, this world wants you to think that it is. They also seem pretty hell-bent making sure you have a taxable locatable "physical body" too. Did you know its a crime to try and discard yours? Well, it *is*. Now, I'm not advocating "suicide"—that's an unnecessary and inadequate way out the bottom—but it seems to me a rather strange law to even have in place—especially when it could be theoretically enforced *with* capital punishment. Now, *there* is a mind-bender for someone with a lot of time to waste. An individual feels so entrapped and bad off about the world and their *Life* that they want to end it all —and the best our society has to offer is to imprison them even deeper into physical systems and suspend them in apathy. *Awesome*. Well—I'm hoping we might be able to do a little better for the Human Condition with our *Cause*.

I'm going to put everyone at ease now and let you in on what we are actually after that is buried under the *personality pro-*

---

110 **misappropriated** : put into use incorrectly; to apply ineffectively or as unintended by design.

gramming inherited from others: its called *"Individuality."* You strip away all the stuff you have accumulated as a filter of experience and purpose that you call "personality" and you discover the *"Individuality"* of the *Alpha Spirit*—the truest "personality" of the *Self*, if you want to call it that. All creative expression and clear thought is a result of *Individuality* and not *personality*. If you want to learn how to write like Shakespeare, then go to school and study Shakespeare. If you want to learn how to write from *Self*, then actualize the creativity that is accessible at these higher levels of *Awareness*, because you sure won't find it buried in one of these so-called "personalities." This example comes up a lot with me, because I am constantly asked about going to school for writing. Personally, I've never understood that. If you need someone to give you a structured outline on how to write or think creatively, well, I'm not sure what to say about that. That is not what we do in *Systemology*, even with all the systematized codification. We are assisting the *Seeker* in actualizing their own creative abilities across the board—not instructing on how they should personally apply them. That involves "individuality"—and it is apparent, at least to me, that "individuality" is presently in short supply among Humans on Earth. You want to know why? *Too much personality.*

I'll give you a perfect example of the problem with "too much personality" going around today: it's called "political correctness"—at least I think that's what they are calling it; *what a stupid term*. It just screams "fragmentation" right in its syntax. So now we are inhibiting *Self* in every type of *expression* that might just piss off some whiny little... Yeah, see even my statement in pointing it out isn't even "politically correct." So, we got that going for us in this world too, because "personalities" are excellent database storage boxes for "feelings" and well, we just can't have people playing with other people's feelings all *willy-nilly*. Even though that's what put us in this position to begin with. And most of that hurt originally came down to us through *imprinting* by some social authoritarian source. Now that we got the population all *frag-*

mented into confusion they go: well, that's enough for now, because it's only up to us "up here" to dispense any conditioned programming to you all, so just, go now and—"sin no more"—uh, do right by each other...somehow—*you know?*—they'll let you know when you flub. Someone will be around shortly to "tell you how it is."

△ △ △ △ △ △ △

We tend to set our sights on the *Awareness* and *actualization* of "spiritual consciousness" as a means toward a greater certainty of our "mental consciousness"—just as we use our existing *Systemology Processing* to treat an elevation of "mental consciousness" by the reduction of emotionally encoded *imprints*. This is why we call our holistic approach to *Life* and the Universe: "*systemology.*" We see how things work *systematically* and we apply this reasoning and certainty to personal management. It is actually all quite simple because it is our basic state. What has happened is that we have complicated things and so the only aspects to all of this that seem complicated are confusing because of the programming that is carried into this—not because this *Systemology* itself is complicated. It's actually easier to understand than any other science. I think many here have found it far easier to grasp than twelve years of "personality indoctrination" in the public school system.

In the process of *defragmenting* frequencies of "thought" and "intellect" maintained at the apex of our *beta-existence*, we are approaching levels that *exceed* what we classify as "intellectual" or the Mind-Systems. There are instances when we *do* naturally tap these states while occupying an *Awareness* in the Physical Universe—and these are instances of high-frequency creative expression at the very least. "Imagination" and "Individuality" are properties of the *Alpha Spirit* that exceed the boundaries of the Physical Universe. They may certainly be realized *in* the Physical Universe, but the "ideas" begin or have a *cause* that is *exterior* to the Physical Universe

—the parts used to express a unique individual *creation* are not just duplicating an existing archetype. We aren't talking about conceiving a "better table" or something. We all know what a "table" is. But that idea was first formed *exterior* to the Physical Universe and then *realized* into *beta-existence* by an intention of *Will*. They aren't just growing ready-built tables on trees or mining them out of the ground. When left to its own accord, no matter how chaotic and random particles may be, "Nature" does not just "manifest tables and chairs" regardless of how long we allow chance to take effect. This is one of the things that has always made the Human Condition on Earth quite suspect from a scientific standpoint—because its existence does not result from either *random chaotic chance* or a purely *natural evolution* that we can be sure of.

Much like the variations found within the physical systems, emotional systems and thought systems, the state we describe as *Self-Honesty* is more of a range within the spiritual systems rather than a specific all-inclusive "point" that is a standard for all individuals. Such would more or less defy the very definition of "individualism." So, what we are really referring to is a "realization" of the higher levels of existence while still maintaining our *beta-existence.* As is demonstrated every step of the way on the *Pathway to Self-Honesty:*

"Realization" *precedes* "Actualization."

Those *Seekers* still working their way through the muck of lower more solid levels of *Awareness* are going to hear this and just think we are splitting hairs with word play. And yet those who are actually working through *Systemology* education and effective steps toward *defragmentation* of "mental consciousness" will understand almost immediately. This applies with all aspects of *creation* and *universe design* higher up the ZU-line, but for our present purposes, it should suffice to say that a *Seeker* is essentially "realizing" *Self*, before they are "actualizing" *Self*—at each point we are *realizing* an "understanding" of *Self,* before we are *actualizing* an "*Awareness*" of *Self*. This should almost go without saying—but we have to say it. And we have to be clear about *realizing* it, so that we can *actualize* its certainty as an axiom of *Systemology Processing*. Therefore, it follows that:

Before you can actualize potential of the Spirit as Self, you must fully realize that you are the Spirit as Self.

Most of the initial work that is emphasized in any mystical, magical or spiritual tradition centers on the establishment and actualization of the "I-AM-Self" before extending attentions beyond. In fact, the Ancient Mystery School—and many of its still existent remnants—focused most heavily on personal development and Self-Actualization before applying it to greater macrocosms[111] of *Self*. When esoteric masters and mystical adepts secretly instructed apprentices in the underground, very little of a mentor's time was dedicated to giving them correspondence lists and rituals and applications that we find so commonly given as standard issue in the "New Age." That information has always been available—and it has always been researched in dusty libraries on an initiate's own time. The lore part is readily discernible from *correspondence charts*, *alchemy tables* and various *kabbalistic models*—such as those preceding our *Standard Model*.

Our *Standard Model* in *Systemology* is actually derived from ancient cuneiform tablets from Mardukite Babylon that describe the inception of systems governing the Human Condition. The system they refer to as the BAB.ILI—meaning the "Gateway to the Stars" or "Pathway of the Gods"—is the direct source of our *Standard Model* and foundation regarding *Systemology*. We are not duplicating an exact reconstruction or attempting to simply revive some ancient tradition verbatim. Even modern *Mardukite Zuism* does not pretend or even bother to do that—because it wouldn't be relevant to our times or workable to the individual in present society—or even very useful for our futurist agenda. It would simply be

---

111 **macrocosmic** : taking examples and system demonstrations at one level and applying them as a larger demonstration of a relatively higher level or unseen dimension.

one more authoritarian standard with no actual effective method of allowing a *Seeker* to experience this truth for themselves.

Basic "*Self-realization*" or "*Self-consciousness*" develops naturally as a result of *Self-Processing*, but it does not happen all at once or even at the same rate for all *Seekers*—it unfolds as individually as the individual themselves, and this uniqueness is a direct result of our *Spirit* and not some artificially created *persona-personality-program* that we use to fill in the gaps of experience in place of true "*Self-Awareness.*" It's kind of funny —I think I've lost count of the number of times I said "Self" in the last thirty seconds, and it reminds me of another semantic trap that a *Seeker* must *defragment*, which is the concept of "Self" and how we are socially conditioned to avoid this realization. We are told that maintaining a heightened state of *Awareness* that maintains *Self* as the "Cosmic Center" is perverted by the physical systems as some kind of negative connotative state of "self-centeredness" only. But, actually:—

ZU is the very spiritual essence of all *Life* in existence and *Self* is a concentrated focal center point of the ZU continuum.

When the *Seeker* is instructed to focus their *attentions* on the physical body throughout various mystical and esoteric exercises, the point of the work is not that the *genetic vehicle* is "oh, so important" beyond anything else we might concentrate on. It always comes back to *Self-directed* "determinism" or "control" when we are developing *Self-Awareness* of "spiritual consciousness." Personal development on a spiritual level and true personal management of *beta-existence* go hand-in-hand. An individual does not have to adopt anything "religious" to be operating from *Spirit*. An individual *is* "Spirit" and so if this can be realized—*truly* realized—then it can be *actualized*. This does not require any particular dogma or belief to work because the whole system is self-proving if you know what you are looking for and what you are looking at. We already know that:

> Most Humans are "asleep at the helm"
> and flying on the auto-pilot of the "RCC."

Its high time we wake them up to a *Crystal Dawn* at this birth of *Homo Novus*.

△ △ △ △ △ △ △

In *Tablets of Destiny*, I focused on the emotional aspects of the *filters* that are used to experience existence from *Self*. Often, this type of restimulation is a fast track to *resurfacing* the *imprinting* and *programming* that is embedded. There is no doubt about that. But it is not always the most appropriate or efficient route of *reducing* the control that these *filters* have over our *Awareness*. When we get a feel for them, we can usually *analyze* them with a little assistance and reduce them in the band of *Thought*. This is actually preferable if a *Seeker* is generating a high enough ZU-frequency in the Mind-Systems to simply "change their mind" about something—to literally *reform* the energy cemented and solidified as *conditioning* and *programming*. When we apply the analytical techniques, we are still *resurfacing* our *filters* of energy and information processing, but we are able to literally *look at* them instead of *looking through* them—and that is a significant point of *Self-Consciousness*.

In view of the fact that we can actually *examine* them from *Self*, in itself, demonstrates that they are not *Self*. It is the "*I-AM-Self*" directed from the very heights of spiritual existence that is doing the actual *looking*. We can go out of *these* bodies very easily—and examine them or project consciousness elsewhere—because they are not *Self*.

We do not ever displace our true *Awareness* out of *Self* to actually see the *Alpha Spirit* because it *is* the *Alpha Spirit* that is doing the real seeing of *things*; the realizing of *knowings*; the actualization of *beingness*—this is all from *Self*. Now, it should be readily obvious by now to anyone who has studied *Mar-*

dukite Systemology material that: we have various control centers engaging systems that ZU interacts with at various levels. But, the uppermost source of *Will* that provides the *Intention* for the other systems to engage *Self-direction* is obviously operating "higher" than—or *exterior* to—the Physical Universe band of KI—and we have now called that the "ACC" or *Alpha Control Center*, which is marked at (7.0) on our *Standard Model*. The energy circuit of existential *Awareness* also runs the other way. The information and sensory processing of environmental energies is passed through all filters and belief systems before returning to the "MCC" and *processed* by *Self* as the "effects" of *Self-directed* "Reality." The *clarity* of the view is directly related to the degree of *fragmentation* or *Awareness* maintained along the ZU-line.

I'm pleased to officially announce that our new *"Self-Help"* book will be titled: "Crystal Clear." The title reflects the exact state that we want to see these crystalline slates[112] of holographic *imprinting, conditioning* and *programming*. Some of you may remember the "prismatic lens" image used to promote the original *"Reality Engineering"* series and its corresponding textbook many years ago. If you weren't around for that—or do not have that book—the same image was made popular from a *Pink Floyd* album. Okay, well, there really is no better tangible model or image of what *Actualized Awareness* really is. It is the projection of a *Clear Light* from *Source* through the pyramidal crystal form of *Self*, and the projection of whatever state that *crystal* is in, as the existence we experience—the array of colors and lights on the wall, so to speak. It's all the same *light*. It's a singular source of pure *light*. And with it we get this pattern array of color—and *how many* colors? Ah—*seven*. Oh, my goodness, well, lookee here: there are *seven* levels of *fragmentation* of *Clear Light* to get back to the *Source*. It's funny, because you know I think the ancient Mardukite Babylonians seemed to have something similar to say about seven levels. I'd have to check though.

---

112 **slate** : a flat surface used for writing on; a chalk-board.

[*Audience laughter.*]

One of the ways in which we are able to use our *mental faculties*—even beyond *resurfacing* or *analyzing* memory and *facets* that we have already experienced—is to take true *knowingness* and perform experiments with *beingness*. This is, apparently, how Einstein discovered so many things, which he describes as "thought experiments." Well, today we might call that "constructive imagination" or even "astral vision"—because the degree of frequency that is drawn from directly relates to what some mystic traditions refer to as the "astral"—which is essentially a proper name for the boundary threshold between a Physical Universe (KI)—*any* Physical Universe, by the way, because this isn't the only one—so, it is a threshold between KI and AN.

Of course, if we are going to be both literal and esoteric about it. The "astral" is a field of *"light."* It is actually a term derived from the Babylonian name for a "deity from the stars" or "a deity of light" called the *Ishtari*, which a person might otherwise just translate as "gods," but it means more than that. In fact, the term more or less replaced the word *Anunnaki*, which was a much older Sumerian concept revisited frequently by the Babylonians. In the beginning, the *Anunnaki* were the "judges" decreeing the "fates" of the Human Condition—yet by the time of the Babylonian systematization, they were essentially "star-beings" and they were able to travel easily back to their "Source" in the Spiritual Universe by *using* this "astral." Now, the ancient common population really didn't have a super handle on metaphysics yet, themselves—that's what the priests and such were for—but they did enjoy a nice healthy indoctrination of *language* for their newly conceived systematized civilization, and so the words relating to the "astral" were attached to the visible "stars" in the sky—just as the names of the "gods" were attached to the visible "planets" that were nearest to us, and which seemed to have some causal relationship on our "fates." You will find this in all the celestial mythologies that came later in any culture.

Now that I have caught you up on the history lesson and semantics—what do we do with this? Well, I am pointing all of this out to give you some background on the energy behind "mental realization"—which is really the "primary thought-intention" that *creates* the "mental image" that we play with in our "Mind's Eye." These images are directed from *Self* as the *Alpha*—which is in direct communication with the *beta* "MCC" of the Mind-Systems. However, our stores of programmed images, and our stores of *imprints*, may also be excited into *being* as a "reactive-response," even if the information contained within them is not actual information being sent from our environment. In essence, when we are not seeing effects of our cause accurately, our determinations become fragmented because the information received is fragmented.

As we imbue more of our *emotional* energy into thought-forms, they become *real*, they become *solid* to us at the very least, and they will affect how we think and what we do. If these are created, played with and reformed at *Will* without becoming a personal energy trap, we are on the right track. Now, the point here is that we can also be *Self-directing* this activity that they call "consciousness," and that is what we should always be doing, so long as *Self* is not inhibited in this effort. It is only when our "efforts" become "unrealized" that they turn back on us as fragmentation. Negative or harmful efforts still stem from some point of *survivalism*—however reactive its point of control may be—but what they result in is actually not productive to creation: the effect is *again*, fragmentation. In some sense, there is a *systematization* of what they call "karma" at work here—and it is entirely governed by Cosmic Law. No intermediaries are technically necessary to administer causal punishment by the Physical Universe.

So, what I would like to introduce to you tonight—and to our growing battery of processing—is not so much a "process" in itself, although we are calling it that for codification sake, but actually a basic command line for existing procedures. It is more of a supplemental procedure that is added to our existing idea of *"Resurfacing"* emotions and *"Recalling"* images that we may *"Analyze"* for information drawn out of the experience. This is essentially no different than projecting *Awareness* "back" in time. The body is staying present, but *Awareness* is directed beyond the present as a means of contacting information that is in the *past*. This information is *energy*—there is some type of energy there to contact for it to be a memory.

Thought energy is fairly light and high-frequency, but when the retrieval of information is becoming the cause of an emotional effect, that means there are *facets* of energy wound up and fragmenting—its blowing up a balloon in our field of vision that contains a message, but this intrusion is not *self-directed*. That is the difference between fragmentation and the information we store as basic thought-memory. This new "supplemental procedure" or command line will be introduced in the *"Crystal Clear"* book as "SP-2B"—as in very literally *"to be"*—and it is a prompt for *Self-Processing* in addition to the standard *"Recall"* and *"Analyze."* Of course, when I tell you the "word" you will just smile—but I'm sure you can see that it opens up the channels of higher frequency energy potential in *Systemology Processing* and is quite applicable and effective in *Piloting*.

Are you ready for the secret word?—The amazing "spell" that can, if effectively applied, dispel all fragmentation of the material world, and nearly every problem facing the Human Condition. That secret word is: "IMAGINE."

△ △ △ △ △ △

Within ancient Sumerian cuneiform vocabulary, we discover a very interesting word for "IMAGINE":

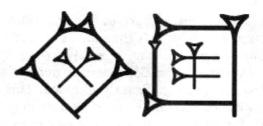

SAG.DAB — "to imagine; conceive of; conjure an idea in Mind."

You and I both know what "Imagine" means—we are calling an image, possibly even of holographic quality, to our Mind's Eye. We are literally "conjuring" it in our Mind's Eye. We are "realizing" it as a potential Reality in the Mind. Now, that doesn't make it automatically "actual" or that suddenly everyone agrees with it and it suddenly "materializes" out of thin air. That's not how this works at *beta* levels where you will always be required to work alongside Cosmic Law and the *way things are*—but as we know from the esoteric axioms of old:

Action follows Thought.

Everything that we can *have*, anything we can *make* or *create*, every physical *action* or activity of *thought* must first be originated and realized from *Self* to ever be actualized as a *Self-determined* cause. This is also an action. There is no way around—unless you're content with just succumbing to be an effect of someone else's cause. That's always an option. But we are assuming that if you are here now, that you are interested in increasing personal management. Okay then, so...

...our efforts and actions would then only be limited to the extent that we can *imagine* and then take *right knowledge* and apply an appropriate *evaluation* of personal *effort*—in light of right experience or Cosmic Law—to *cause* an *effect* in accordance with *Will*.

And is that not what "magic" is?—Or, the amazing abilities we read about in stories of very famous and very powerful Self-

Actualized "avatars" back there in ancient times? What has actually changed in all of this time for us concerning our inherent ability—except that the *Awareness* and *Light* maintained by the Human Condition in normal society has been successfully and significantly dimmed. This is not the *end* for us—we most certainly can do something about it in this lifetime. But many are still yet *unaware* of even this fact: that it has even happened; or that there is a remedy. That remedy is a spiritual evolution just as much as it will be an intellectual one. As in the past, the former will go extinct, replaced by *Homo Novus*, which functions with a purpose previously unrealized.

When I first started developing early work for *Systemology* decades ago, the intention behind it was not a therapy and it was not even a literal revival of ancient Mesopotamian tradition. The intention was the future of NexGen Humanity and spiritual evolution as an *alternative* to the "New Age" and other approaches. Do you get that? It's not a "New Age alternative" its an "alternative to New Age" and in the late 1990's, I started referring to this as the coming "Crystal Age" or "Crystal Dawn." Many of you know that back in those days, and when I was gathering information for "*Arcanum: The Great Magical Arcanum*" anthology, I was mostly interested with bridging traditional occultism and mysticism with psychology and quantum physics. That provided a headache I didn't quite expect—not because *I personally* didn't understand it, but because I was at a loss for the proper *vocabulary* and *media* to communicate anything of it directly. Until quite recently, this was once a stumbling block for underground development of *Mardukite Systemology*—it simply required much more time in experimentation, research and discovery to systematize a workable methodology. We have that now.

Before this evening, our education and processing has been purely in direct *reduction*, or the direct *analytical evaluation* of the *past*. This is an important first step in creative processing: the freeing up and gathering of viable resources. It is often difficult to create something new when there is a whole mess of stuff in the way. So that is all we are doing in the beginning steps of the *Pathway to Self-Honesty*: clearing the way forward and lightening the load we are to carry with us on the journey. That is very important because the journey forward is about as forgiving as Nature is. These are the steps that are too often missed in other methods of personal development and why so many individuals that have set out on some former grandiose "self-help routine" or "mystical initiation service" have not found the results that were expected.

So many Humans are out there crawling, a few are up on their feet stumbling about—but the goal right now is to get you to walk and then to fly. You got a guy that *imagines* he can just fly all right just because he can imagine it. Well, he has a personal sense of this *Reality*—but others have imagined it too and they have evaluated it with greater accuracy because they know about planes and physics and birds and perhaps some ancient forgotten levitation exercise or such and they say, "why, yes you can fly, but you will need to increase your Reality on it"—you know?—bring it up to analytical levels and really see this thing through. But, you realize this guy is fragmented pretty bad, and he has the image in his head that he can just fly—and the funny thing is, he might even have a valid picture in his head from somewhere, but it has a hold on him in the *past* and he goes and jumps off a building as he is and says, "yes, see, the feeling of weightlessness and the air against my...." *crash!* Yeah, so what is in the mind of a person like that? We don't think about it—because we don't think about it; we certainly don't go around thinking like that.

That's a serious case of personal fragmentation, right? It would seem like a funny joke, but we all know that this is the case out there with some individuals. We tend to just ignore it unless they fall on our car or something. But that is the point: a highly *fragmented* individual is potentially dangerous; to themselves; to others; to the community. But we have never—as a society—actually managed much headway in this department. And we know the true *Self* is perfect as *Alpha Spirit*. So, we don't really need much further evidence to convince us of the fact of *conditioning* and *imprinting* on the Human Condition; what it can do and where it leads, and so forth.

[*Audience Voice: "Didn't you say that maybe that guy was right though? He was valid in this action?"*]

Did I say that? No... Oh, well, he may have had a valid picture in his head. We aren't going to get too deep into that at this point because some of these possible sources move into areas that—well, frankly—I'm not ready to communicate yet within the confines of our current subject. The picture might be from another lifetime. The nuts and bolts of this I can tell you now: we have found in nearly every instance of *intensive processing* that there is a combination of programming taking place within the Human Condition, from two sources:

   a) the genetic memory[113] of the genetic *beta*-organism; and

   b) the spiritual memory of the *Alpha Spirit*.

There is the obvious third source, being the memory-experience accumulated in *this* lifetime—and again, this is going back to a lot of deep spiritual and metaphysical thinking (even *Alpha* level thinking, if you will) that we are very well on our way to systematizing more accurately. Because we are also talking about *past lives* of *Self* versus the *genetic coding* of the organism that *Self* is using. That's a bit of a mess to sort out, I can assure you. But it is no less the case. We also find that there are "artificial imprints" or *implants* that may be in-

---

113 **genetic memory** : the evolutionary, cellular and genetic (DNA) "memory" encoded into a *genetic vehicle* or *living organism* during its progression and duplication (reproduction) over millions (or billions) of years on Earth; in *NexGen Systemology*—the past-life Earth-memory carried in the genetic makeup of an organism (*genetic vehicle*) that is *independent of any* actual "spiritual memory" maintained by the *Alpha Spirit* themselves, from its own previous lifetimes on Earth and elsewhere using other *genetic vehicles* with no direct genetic connection to the current physical form in use.

stalled that contribute to our sense of "Identity" at each end of the spectrum. And I mean whole scenarios and such, huge packages—more than what you would call "personality" although it lends itself to that as well. Nothing is independent really. You get a lot of concepts about "*karma*" from this over in the Eastern World. Here, they talk about the "Law of Attraction" and such in American New Thought. Former communications of these ideas have all surrounded a basic observation of the same phenomenon, the same *Spiritual Life Energy* that we refer to as ZU, and the same interactions of Cosmic Law—all of which were realized and relayed to the extent that the understanding and vocabulary provided for. Yet, with so many other advancements—external technological advancements—these other spiritual applications have not received the same degree of attention and effective objective results over the past century. If they had, I think we would be seeing a vastly different result right now on planet Earth.

Most of the time we are taught that *psychotic* behavior is just random nonsense emerging out of nowhere—I used to be one of those indoctrinated "psych-majors" in college—but it would seem that there is always a reason, even if the source is *fragmented*. I mean: it would have to be *fragmented*, right? Because the Source is either *Self* or its not and we are the *effect*. There is not a lot of gray area in between there. It's like a light switch. *We* "grade" it in *Systemology* for our own personal convenience in managing some structure of development, but its kind of obvious that we are either *Self-Honest*, actualizing a state of "*spiritual consciousness*" or we are not. Whenever we are not, we are not completely out of the loop, but we are not "at the top" so to speak—not at the "top of our *Game*." And, there aren't really any consolation prizes in this *Game* other than perhaps a sense of personal satisfaction as you work your way up.

But this is a *Game of Destiny*—the destination is what we are after. That requires personal *defragmentation* using every worthwhile means that can be applied to these *Processes* that show definitive significant results across the board. That has required a lot of checks and balances to get this under way, but I am quite confident now that we are very well on our way to successfully and completely charting the total *Pathway to Self-Honesty*. I hope you are enjoying the journey. Thank you.

UNIT FOUR
SELF DISCOVERY

THE DISCOVERY OF SELF-CONSCIOUSNESS

Each and every one of us has the ability to *incite*[114] or *dissolve* "creation" with our *Attention*—our focused application of *Self-directed Awareness*. We may do this many times a day: we form an intention in our mind, create an entire "mental image" and then erase it. This is the innate ability of *Self* to "imagine." And as we know from one of the more intellectual chiefs of the modern age: "Imagination is more important than knowledge" but this is only part of the statement, which is an important axiom of *NexGen Systemology*, as stated by Albert Einstein:—

*"Imagination is more important than knowledge.
For knowledge is limited to all we now know and understand,
while imagination embraces the entire world,
and all there ever will be to know and understand."*

"Imagination" is the highest faculty of the Mind-System typically experienced in *beta-existence* and perhaps better stated: it is a property of the *Spirit*, much like "individuality" is a property of the *Spirit*. It is "Imagination" that allows the "individual" the ability to *Self-direct* focused *Awareness* on *anything* at *Will*—not just what has already happened, what is happening, or what is present in the immediate environment. The *Actualized Self*—as the ultimate "Reality Technician"—has the ability to imagine *any* "image" or "energy" into the present—and apply additional *Will* to transform and manifest its "Reality." We consciously are doing this all of the time, in everything we think, do and feel—but it is the level of *certainty* and *Awareness* that determines just to what extent we are able to "actualize" our "realizations."

Some individuals are certain of little more than *Self-directed* ability to tie their shoes and put one foot in front of the other. Total *Self-control* of the *genetic vehicle* operated by the *Self* for *beta* is a necessary step for total *Self-Actualization* in *this* lifetime. This basic truth is evident from our research and ex-

---

114 **incite** : to urge on; instigate; prove or stimulate into action.

perimentation and what is documented by six millennium of methods and techniques developed from the Ancient Mystery School. An individual will, however—once reaching the levels of "mental consciousness"—become *Aware* of the limitations of *beta-knowledge* as applied to the fullest extent of expression —*Actualization* and *Beingness*—and they will usually approach the problem from one of two ways:

a) <u>add</u> an additional "external-artificial layer" or an "upgrade" for the *beta-existence* to overcome an immediately perceived barrier/limitation; or

b) <u>remove</u> an "artificial layer" or "uncover internal-innate potentials" of the *Alpha Spirit* that are already blotted out by previous *beta* upgrades.

Adept mystics and esoteric spiritual masters throughout the ages have put forth a variety of lessons and practices to their apprentices that reinforce *Alpha control* of the physical body. These exercises are intended to strengthen a *Seeker's* resolve or certainty in *Self-direction* of the physical body—they are not a glorification or emphasis of *Awareness* onto the physical body itself. Essentially, if we are to be incredibly accurate concerning the application of *Self-directed Awareness*, the purpose of so fixedly reinforcing the *control* of the body by direct *Awareness* is to actually maintain a clear enough channel throughout the ZU-line affecting *beta-existence* that the *Self* does not require a focused *Awareness* on the physical body at all. Technically and literally this is a necessary condition for all "ecstatic" states that are *exterior* to the Physical Universe and *beta*-range—above (4.0) on the ZU-line.

SELF-REALIZATION OF THE "LIGHT-BODY"

The term "*dis*-covering" was found preferable to "*un*-covering"—yet the later is more accurately representing the sentiment concerning the "unfoldment" of the "I-AM" Spiritual Self or *Alpha-Spirit*. It is only a "discovery" because we are experiencing it *again* with *new eyes*—but the keyword here is "*again.*" It is actually an "uncovering" of *Self* that already *is*. We aren't "discovering" it because we *are* it. But we *are* shedding artificial attachments used to experience *Self*, much like we might peel away layers of clothing from the physical body. The present author has made numerous attempts—in every genre previously demonstrated in the *Mardukite Research Library*—to emphasize significance of one key aspect appearing throughout all original source materials based on the Ancient Mystery School:

—*an Actualized Realization of the "Light-Body."*

This quality—a realization of operating from the *Alpha Spirit* or truest "I-AM-*Self*"—is the primary thought-wave necessary for actualization of *Self* as *Clear Light*, its uppermost spiritual form that is superior to all *beta*-expression as *Source* or *Cause*. The increase of *Awareness* is demonstrated by techniques formerly described by mystics as the "Right Way"—based on an understanding of *Cosmic Law*. We have simply *systematized* the same, drawing from a much greater wide-angle view concerning *6,000 years* of written records and practical systems descended from the Ancient Mystery School. Much of this we have discarded along the way—meaning that we have kept for *Systemology* only that which was *actually useful* and found effectively workable "across the boards" for this "21st Century New Thought" applied spiritual philosophy illuminating the *Pathway to Self-Honesty*.

There are many esoteric exercises that appear throughout history under the titles of "Body of Light" and "Astral Body" and "Spirit Body"—all of which are descriptions for the *substance* or *spiritual matter form* that our actual spiritual *Identity* maintains as *Self*. It is a "body"—in the sense that it has a form and its own system of organs and motors that exist in the space-time of the Spiritual Universe (AN)—composed of spiritual energy and matter. Such is not directly detectable in the Physical Universe (KI) except where it comes to this thing we call "Life"—and of which we cannot measure directly, only "observed changes" as they *apply to* the Physical Universe.

The Light-Body is essentially the emanation that results from the "spiritual consciousness" of the *Actualized Self* or *Alpha Spirit* as the epicenter of a spiritual manifestation—substance, action and consciousness (*ZU-continuum*) in the Spiritual Universe (AN). The Light-Body is a field of energy generated by *Alpha Awareness* activity that provides spiritual structure to the "I" form we call *Self*. It is this Light-Body that commandeers the *genetic vehicle* directly—and most have realized that the humanoid form taken or preferred is roughly equivalent to the shape of the field-form emanating from *Self* as the Light-Body. The relative energetic spatial mass of the Light-Body is approximately *three feet*[*] above and around all sides of the physical body—which some mystics have referred to in the past as the *"aura."*

—RECALL a moment you heard someone say the word: *Aura*.

—RECALL the first time you heard someone say the word: *Aura*.

—RECALL an instance when you might have seen someone's *Aura*.

—RECALL an instance when you might have sensed someone's *Aura*.

—RECALL a time when you might have seen your own *Aura*.

—RECALL a time when you might have sensed your own *Aura*.

—IMAGINE your physical body is enshrouded in a *sphere of light*.

---

[*] The relative dimensional phase-distance currently used for spiritual-portal experiments by the "*Mardukite Research Organization*" is approximately: *2.72 feet*, or *32.64 inches*, or *82.91 centimeters*. A wavelength of *.8291 meters* converts to *361.588* or *360.588* megahertz (or $361*10^6$ hz). This carries a photon energy wave on the (EM) electromagnetic spectrum of 1.495 microelektron-volts (or $1.5*10^{-6}$ eV)—an (EM) energy value of $239.591*10^{-27}$ J. This is also approximately equivalent to the frequency threshold distinguishing "UHF microwaves" from "VHF radio waves."

—RECALL a moment when *you* realized *you* were a *spirit* controlling a *physical body*.

—RECALL the first time *you* realized *you* were a *spirit* controlling a *physical body*.

—RECALL a moment you heard someone say: *Man is an Animal*.

—RECALL the first time you heard someone say: *Man is an Animal*.

—RECALL an instance when you heard or read the phrase: *Human Animal*.

—RECALL an instance when you watched someone "act like an *animal*."

—RECALL a time when you might have "acted like an *animal*."

—IMAGINE your physical body is enshrouded in a *sphere of light*.

—RECALL a moment when you realized you were *outside of your body*.

When we focus our concentrated attentions on our "realizations" they become "actual"—and this applies to aspects of all *Awareness* levels, both *Spiritual (Alpha)* and *Physical (Beta)*. *Realization*—and subsequent *Actualization*—of the Light–Body in *Awareness* is an individual effort that is not as much prompted by a "*process*" specifically, but through a combined elevation of consciousness activity on the ZU-line using *processing*. "Imagery functions" of higher frequency "Alpha Thought" (6.0)—*Imagination*—are utilized to "*realize*" certain energies in the Mind's Eye (4.0) from WILL (5.0), but it should be well understood that these energies are no less *real*—from an objective point-of-view—even when they are not being "realized" in *Awareness* enough to be treated as an "*actuality*" in *beta-existence*. We know the Sun exists, even when we do not see it directly. We *realize* it is in existence even when it is not *actually* there. And by "actually there," we mean of course, that it is not presently manifest locally in our vicinity.

Yet, someone operating in the band of "mental consciousness" will *realize* that an *actual* effect of its light is illuminating the moon! And since we know that personal *Awareness* influences the external "world of things" to the extent that *Awareness* is actualized to manage a certain "sphere of influence" and responsibility, we find that:

"Increasing Realization" = "Increased Actualization."

In previous lesson-chapters the *Seeker* is prompted to practice an extroversion of focused *Awareness* on physical objects, forms and spots in physical space. When actualizing the Light–Body, the same practice is introverted to place total focused *Awareness* on the internal points of the physical body. This also reinforces that the *genetic vehicle* is directed as an effect from *Self* as the causal source. Because of the type of ZU energy attached to *Attention-Awareness*, the focused concentration of *Awareness* on the body is not an effort to become literally *more* Aware of the "genetic vehicle" itself, but to increase the communication and control of ZU energy between *Self* as *Alpha Spirit* and the physical body as a *genetic vehicle* for *beta-existence*. This actually *reduces* how "Aware" the *Self* has to be on the body, but it is not removed from the equation, such as when the RCC kicks out our *Awareness*. Instead it is operating *Cause* from exceedingly higher "seats of consciousness" than the MCC, which is now acting as a direct relay point, rather than being treated as the "height" of our potential *Awareness*. This is possible only when the ZU-line is "in phase" with Self so that there is no turbulence at lower frequency levels.

The more an individual actively fixes *Awareness* to lower points of turbulence, the less likely they are to achieve the higher levels. This is not intended to seem cryptic, misleading or excluding. It is simply the way the systems work. As mentioned at the very beginning of this course, true progress on the *Pathway to Self-Honesty* cannot be "cheated"—but it may be accelerated by the proper dedication and application of "right knowledge" and "right actions" just as much as it can be decelerated and thwarted by ignorance or neglect. The effort is a matter of *Self-determination*—and for the first time a structured organization exists that provides the tools without standing in the way as an authoritarian intermediary to the relationship between the *Seeker* and *Infinity*.

ACTUALIZATION OF THE "LIGHT-BODY"

A *Seeker* may safely and easily practice "*Actualizing the Light-Body.*" Rather than something that is found within the form of the body, the Light-Body may be best realized as a *Self-Awareness* that enshrouds the physical form. It maintains energetic contact with points all throughout the physical body —and it may even project its *Awareness* to any one of them— but its point of view is not "inside" of the body or behind the eyes. There are potential contact points all throughout the body which may each be reached in turn for practice. There are *three* specific "energetic control centers" identifiable along the ZU-line—each able to be *defragmented* with various levels of *processing* that will in turn enable higher control centers to naturally govern subordinate systems to optimum efficiency.[115] Energy will naturally seek optimization to the extent that its channels are clear.

> —There is a primal energy center for the *genetic vehicle* itself that resides around the "solar plexus," and we may easily identify it as the RCC at (2.0) on the ZU-line.
>
> —There is the matter of the "Third Eye" just behind the forehead that corresponds to the MCC energetic contact point at (4.0).
>
> —The actual "Spiritual Light-Body" of the *Alpha Spirit*, we can only use relative space to align to our Standard Model, but we have suggested that this "ZU-aura" enshrouds a living organism and extends at a wavelength of approximately 2.72 (*or three*) feet out from it. The ext-

---

115 **optimum** : the most favorable conditions for the best result; the great degree of result under specific conditions.

ent we may distinguish an individualized *Alpha Self* is the ACC plotted at (7.0).

We can project *Self-Awareness* from any of these levels and perceive conscious activity at any degree. It is quite obvious, however, that as higher control centers are realized, the stronger, higher-frequency, ZU energy is actualized. The huge secret here—which we will share—is that these alternate *states of beingness* may be realized through practice without first having consciously experienced them. This is actually necessary to actualize *Beingness*. It must be first imagined and then *willed* into form—hence "practice makes perfect." It is a perfected actualization of *Beingness* we seek—and it must first conditionally be realized to be actualized. Thus—*Systemology Processing* is practiced as a *means* to perfecting the same. The basic principles remain unchanged regardless of the level or Grade on the *Pathway*, yet at each point the *Seeker* is bring a new level of realization to the *Game*, using a higher realized "*kNow-ledge*" of personal certainty toward further attainment of *Self-Actualization*.

Once a *Seeker* is able to clear up enough personal energy channels from approximately (3.0) and *below* (on the *Standard Model/ZU-line*), actualized *beta-Awareness* has reached 50%. Surprisingly, this is well-above normal in our modern society—perhaps even well above the scores you are currently maintaining using the *Systemology BAT tests*. It is at this point that we reach a significant shift in consciousness flow between:

a) the personal emphasis and focus of attention on issues currently faced as *present conditions* of Reality (*2.1-3.0*); and

b) the personal expression of basic ability to consider and apply effort toward *future conditions* of Reality (*3.1-3.5*).

It is here, combining this information with former, that we may more easily observe that bands (or ranges/levels) defining particular relative states are "closer" in proximity or energetic affinity[116] but represent greater amounts of energy —or if applied to a *physical map* of a *snakelike-path*, there is more mass within a shorter increment of increase than at lower degrees of vibration. For example:

—There is a wide band of emotional activity between (0.1) and (2.0)—that is nearly two complete incremental levels on our *Model* dedicated to physical matter and storage of past energy.

—Compare this to the "first order of analytical reasoning" of the *present* between (2.1) and (3.0), which represents the singular standard-issue complete step that most people confine all thought potential to.

And by a society that sets it sights on (3.0) as the maximum range—the laws of averages, conservation and affluence demonstrate—it only yields a population that treats between (2.1) and (2.5) as "normal."

During the course of a *Seeker* practicing realizations of the Light-Body, any number of experiences, sensations and phenomenon are possible. No statements or suggestions will be put forth at this time regarding any specific results, so that we are not leading the *Seeker* with any "positive suggestion" toward artificially producing an effect. In fact, enough subjective literature is already existent regarding spirituality, mysticism and religion over the past few thousands of years that someone really curious on the experiences of others can simply look around. We are mainly concerned with the actualization of a *Seeker's* own potentials here.

The *Seeker* will also—while on the journey of the *Pathway*—spend a considerable amount of time directing personal *processing* toward the actualized *Self*. This is prompted *to be* done "from *Spirit*," "from *Alpha*" or "from *Source*" as the individual

---

116 **affinity** : the apparent and energetic *relationship* between substances or bodies; the degree of *attraction* or repulsion between things based on natural forces; the *similitude* of frequencies or waveforms; the degree of *interconnection* between systems.

prefers to assign the label. Any results simply yield themselves as a natural result of consistent reinforced realization of *Self* as *Alpha Spirit* and not simply the "body" or even the "Mind"—as we tend to be conditioned to believe. In fact, we may very well discover that reclaiming the power of "spiritual consciousness" is simply the untouched key to all of the so called "supernatural" phenomenon that modern society has otherwise failed to readily understand and tap using traditional programming of the Human Condition.

There *are* very ancient techniques designed to blatantly reinforce the Light-Body with *Awareness*—and there are, in fact, so many various esoteric examples that an entire volume of materials could be devoted to little else. That is, of course, not the present intent. Therefore this chapter-lesson is intended to provide education and descriptions of use, rather than a rigid set of rules and steps that arguably "must be" followed in order to achieve a specific result. Keep in mind that what a *Seeker* is working toward equates to the natural, basic, true state of *Self* via removal of artificial garb. Any of these methods and techniques provided in this volume—or throughout *Systemology*—require a certain degree of "personalization" in order to be universally effective.

Raising personal *Awareness* to the Light-Body may be accomplished through any one of several different direct practices, or some combination of the same. Although there are many religious or esoteric examples available, a *Seeker* does not need to adopt any particular *belief* about such things in order to make use of effective tools, or transform existing ideas. Suggestions regarding attempts toward *Self-Actualization* may be located at the heart of nearly all spiritual philosophies for thousands of years. What we are most concerned with are the methods that are effective and which will not contribute to any additional *aberrated fragmentation*.

Some methods—discovered on account of more than two decades of personal research by the present author—operate on a principle of starting with an *Awareness* in the feet or the head or some extremity and carrying that *point* to other areas of the body. Other methods indicate targeting specific points of the body with *Awareness* and allowing this to radiate outward from each epicenter. There are esoteric methods that have assigned various names of *Divinity* or *Spiritual Energy* to each of these parts—such as in the *kabbalahs* of the *Ancient Near East*, where each part of the body was assigned the name of a "spirit" that controlled it. We cannot be absolutely certain that there are not other spiritual entities that also seek to gain *Alpha* control of our *genetic vehicle*—but the one thing we can do is raise the level of certainty that *Self* is the *Alpha Spirit* operating in *beta-existence* as "I" without interference, dissonance[117] or other *fragmentation*. These provide a point of reference to the *Seeker* regarding the operations employed to this effect, even if they are not utilized directly.

Of the dozens of examples personally explored and experimented with to gauge varying degrees of effectiveness based on the *process* alone—and not as a result of the skills of the operator—there were several key aspects that formed patterns between any written instructions and their experienced effects. As a result, methods of standard processing specifically toward the conscious "realization" of *Self* as the *Light-Body* or *Alpha Spirit* were actually developed following basic *Systemology Processing* formulas. However, even without taking up a standard routine, the following aspects are common variations proven successful, and which may certainly be applied as personally desired to a method of a *Seeker's* own design. The most effective methods of *Light-Body* work included:—

- Operating a process in a *relaxed* state, in a relaxed environment free of distracting influences, sounds and other *facets*.
- Imagining/Sensing any directed focused/concentrated

---

117 **dissonance** : discordance; out of step; out of phase; disharmonious; the "differential" between the way things are and the way things are experienced.

energy of *Awareness* on the body as "*Light*"—especially "*Clear Light.*"

• *Awareness* is first directed as an *attention* that acknowledges the whole physical form as a complete system before focusing any energy inward or on any specific parts of the whole.

• Whether focusing *Light* on specific points of the body or directing the passage of *Light* along a course, *Awareness* is brought to either the *feet* or the *head* at the start of the exercise—and then it is sent toward the other direction, hitting all key points and extremities in sequence along the way.

• If a *Seeker* finds difficulty gaining a sense or focus of *Awareness* at various points of the body by only directing the Mind, the exercise is supplemented with practice of first focusing the *Awareness* attention and then actually "touching" that point of the physical body with your finger tip. This increases connectivity of *Self-control* in each specific spot.

• When pertaining to parts in pairs—feet, legs, arms, hands, &tc.—a *Seeker* may alternate focus between the right and left with equal effort and speed. For example: the big toe on the right foot and then the big toe on the left foot. Those finding difficulty with this may practice with bringing a specific side into realization before the other—such as the entire left foot before the right foot. The idea is to balance control of both sides of the body.

• Some exercises direct an introversion of *Awareness*, prompting "realization" of *Self* as residing "within" and enveloped by the systems of the physical body—whereas others direct an extroversion of *Awareness*, prompting "realization" of *Self* as shrouding and "enveloping" the systems of the physical body.

• Effective techniques place an *Awareness* on a "physical form" only as a *Self-directed* practice of placing selective attention on "spiritual forms"—which naturally promotes increased energy and *Self-control* at intermediary levels of "mental consciousness."

SYSTEMATIZATION OF SELF-CONSCIOUSNESS

The relationship between *Self* and the management of *Self*, *Reality* and the *Universe* is best related to the *ZU-continuum* on the *Standard Model*—which also coincides with what we subjectively refer to as "Spheres of Influence" regarding the "reach" of our *Self*, and objectively as "Circles of Existence" concerning manifestation in *Universes*. This *ZU-continuum* exists as an "actuality" even when an individual is not "realizing" it. However, a *Seeker* can only make "actual" use of *Self-directed* communications of energy along the ZU-line to the limits that this *Awareness* is being "realized" as "actual." If this is not understood clearly, the *Seeker* should refer back to previous sections regarding *"realization-versus-actualization."*

> SELF-DETERMINISM & SELF-CONSCIOUSNESS
> (SYSTEMS INTERACTION ON THE ZU-LINE)
> 8.0—Absolute Beingness—"Infinity" (-AN-)
> 7.0—Alpha-Spirit "I-AM"—"Cause" (ACC)
> 6.0—Games & Logics—"Alpha-Thought"
> 5.0—Self-Directed *Will*—"Intent" (*to MCC*)
> 4.0—Evaluation by "Beta-Thought" (*MCC*)
> 3.0—Self-Directed *Effort*—"Extent" (*to RCC*)
> 2.0—Emotive/Physical "Re/Action" (*RCC*)
> 1.0—Beta-Existence/*Gen. Body*—"Effect"
> 0.0—Physical Matter—"Continuity" (-KI-)

The *Seeker* should take some time now and consider the relationship between systems on the ZU-line. These are the systems, states and levels of ZU energy in operation as "consciousness" for any spiritual *Identity*—or "I." These systems, for the individual *Self*, also interact with other fields of similar energy that are contacted in the environment—whether physical *or* spiritual. We tend to only consider the most rigid physical levels of personal interaction with environment, but as we have explained: there are also "emotional," "mental" and "spiritual" degrees of ZU-energy interaction, which also compose their own environmental-plane or level. We simply refer to these as *"systems"* in *Systemology*.

As is indicated on any of our *models*, *scales* or the basic design of a *pyramid structure*—"supreme power" comes *downward* (from the direction of AN) as "cause" to (the direction of KI) as "effect." The esoteric philosophies of causality, sequence and *cosmic ordering* have put forth efforts to demonstrate these same principles for thousands of years—and yet the true understanding was never *systematized* for the written word. For example, with this model you can demonstrate definitively in clear language how the true *Alpha* "I" solidifies "thought" and "will" into existence. That's new. Such "mystery" was formerly left to an initiate to simply experiment with and sort out as *actualization* for themselves—*or* not—and we certainly can look around and see how far *that mode* of operations has progressed us (as a generalization).

A *Seeker* that is making use of *Systemology Processing* as a means of development for the *Pathway to Self-Honesty* applies personal attention to make incremental increases in personal certainty regarding *Self* as *Cause*.

> The greater our realization of *Self*, the greater the ability to *Self-direct* ZU as *Cause*—which is the ultimate function of all *Spiritual Life*.

The infinite potential inherent in our "attention"—our *focus* of *Awareness*—clearly demonstrates this to be the case. The lower the degree that an individual fixes *Awareness*, the more of an *Effect* (and less of a *Cause*) experienced—which in turn continues to lower *Awareness* or keep it bound up in the *imprints* and *images* occupying it there. This is how our *Awareness* and potential of *Self* is operating all of the time.

When an individual is primarily "reactive," they are putting themselves down in the bands (range/levels) of *Effect*. This means that as opposed to *Self-directing Self-Awareness* to "act" and "create" and "Cause," the individual's personal integrity as *Self* is being impinged upon by the environment and external influences. The irony is, these forces are always in existence at these levels—and it is only when an individual succumbs to fixing *Awareness* there for reactive mechanisms that they suddenly feel "overwhelmed" by the "pressures of the world." We can assure you, dear *Seeker*, these pressures and forces and dross[118] of material existence and physical gravity of KI is very much always extant.[119] As the individual lowers in personal ZU energy, "internal pressure" seem insufficient to keep the "external pressure" from almost literally caving the individual's physical form in. Consider, perhaps that there are "immune levels" necessary to maintain optimum *emotional*, *psychological* and *spiritual* well-being.

The chart-table (*listed above*) demonstrating "Self-Consciousness System Interaction on the ZU-line" is arranged with an intention to display the total *ZU-continuum* in this regard, with particular attention to how *Cause* becomes *Effect*—or how *Alpha Thought* (6.0) of the *Alpha-Spirit* (7.0) is processed and becomes actions and experiences of *beta-existence*. However, as with all *systematic* demonstrations that are correctly integrated/calculated for *Systemology*, this pathway of ZU also circuits energy and a communication of system interactions in the "other" direction: from the *experience of effect* back up to "spiritual knowingness" of the *Alpha/Self*. It also assists in understanding (and demonstrating) turbulent flows on the ZU-line of *fragmented* information, sent from "reactive programming" or essentially any point of consciousness processing along the way.

This newly perfected and workable holistic "Standard

---

118 **dross** : prime material; specifically waste-matter or refuse; the discarded remains collected together.
119 **extant** : in existence; existing.

Model" for *Systemology* may be used to effectively understand the operation of—and clear out *fragmentation* from—the operation of personal energetic communication channels of ZU.

The "mental consciousness" that is experienced in day-to-day life, if *defragmented*, is 100% *Self-Directed* by the "I" as *Alpha Spirit*. This is to say—using our new table/understanding—that Self-directed WILL (5.0) of the *Spirit* (*Self*) is communicated as ZU-energy on an *Identity-continuum* to the MCC (4.0) as 100% *beta-Awareness*. This is the precise reason we evaluate the threshold between Physical–Spiritual, *Alpha–Beta*, &tc. at (4.0)—and also our scales, tests and other mathematically coordinated logic. All of these are relayed in *Systemology* exactly as they are because it is the only way in which the holistic picture actually works from the highest echelon of "Games" and "logic."*

We have described *Awareness* as "*Life*" or "*Spiritual Life*" using an ancient Sumerian word for "spiritual consciousness, knowingness and beingness" as ZU. In it's purest state as Infinity (8.0)—in the direction of AN—we consider this expression as AB.ZU or "Absolute Cause"—which is about as close to defining any specific quality of "God" as we can in *Systemology*. In fact, any and all true qualities that may be semantically attached to a concept of the "Absolute Divine" would pertain very specifically to qualities of Infinity (8.0). Even the original primary or first "Alpha Thought" *to be* "an individual spiritual Identity" as *Self* (7.0) is technically, from the perspective of Infinity, a form of separations and fragmentation, in and of itself.

---

* The educational subject of "*Games, Systems & Logic*" is treated in higher *Grade* materials of *NexGen Systemology*, which is gauged to directly reflect the knowledge-base and *Awareness* demonstrated as "levels" on the *Standard Model, Total Awareness Chart* and other interrelated presentations of *Systemology*—on which the "*Systems Interactions of Self-Consciousness Table*" (given in this knowledge unit) is also directly derived from.

Spiritual *extropy*[120] causes spiritual matter and spiritual energy of *Self* to gravitate toward the direction of AN— thereby returning the Source; spiritual Source—just as physical matter and physical energy of a *genetic vehicle* is pulled by *entropy*[121] to gravitate toward the direction of KI as inert matter.

This progression of existence is quite logical in accordance with Cosmic Law—but so is the *Self-determinism* to direct energy by *Thought, Will-Intention* and *Effort*.

CONSIDERATIONS: THOUGHT, WILL & EFFORT

Personal management—and the *Processing* used to rehabilitate it—are possible because of an amazing faculty of the *Self*, we call *"consideration."*[122] There is a tremendous similarity between the manner in which *Self-Consciousness* is *"actualized"* and the effectiveness of *processing*, the application of a *process*, and the efforts applied (and duration) for running *processes*. All of this falls within a very sweeping activity of total consciousness—all consciousness systems—which we may sum up with the a very basic concept of *"consideration."*

There is the *right effort* that may be applied to yield the most optimum efficiency of results—and then there is everything else that falls short or even exceeds this, which is usually not efficient or effective in yielding the desired results or effect.

---

120 **extropy** : in *NexGen Systemology*—the reduction of organized spiritual systems back into a singularity of Infinity when their integrity is measured against space over time.
121 **entropy** : the reduction of organized physical systems back into chaos-continuity when their integrity is measured against space over time.
122 **consideration** : careful analytical reflection of all aspects; deliberation; evaluation of facts and importance of certain facts; thorough examination of all aspects related to, or important for, making a decision; the analysis of consequences and estimation of significance when making decisions.

The faculty that enables proper gauging of *effort* toward "manifestation of things" or "actualization of beingness" is what we call "*consideration.*"

> Our ability to properly analyze, evaluate, realize, actualize—and *get "effect-ive" results*—is a matter of "*right consideration.*"

We should hardly need to apply some dogmatic or moral imperative to the "rightness" or "wrongness" of our efforts when it is *Self-Honestly* evident by the *results*.

—RECALL an instance when you heard someone say the word: *Effort*.

—RECALL a moment when you used the word: *Effort*.

—RECALL a time when you applied *Effort* to accomplish something, and you were successful.

—RECALL a time when you applied *Effort* to accomplish something, and you were unsuccessful.

—RECALL an instance when someone said to you: *Give it more Effort*.

—RECALL an instance when you said to someone: *Give it more Effort*.

—RECALL an instance when someone said to someone else to: *Give it more Effort*.

—RECALL an instance when someone said to you: *You're trying too hard*.

—RECALL an instance when you said to someone: *You're trying too hard*.

—RECALL an instance when someone said to someone else: *You're trying too hard*.

—RECALL an event when you witnessed someone applying *Effort* to accomplish something, and they were successful.

—RECALL an event when you witnessed someone applying *Effort* to accomplish something, and they were unsuccessful.

—RECALL a time when you applied *Effort* and someone applied *Effort* against you.

—RECALL a time when you applied *Effort* against the *Effort* of someone else, and you were effective.

—RECALL a time when you applied *Effort* against the *Effort* of someone else, and you were ineffective.

—RECALL a moment when you witnessed someone applying *Effort* against the *Effort* of someone else, and they were effective.

—RECALL a moment when you witnessed someone applying *Effort* against the *Effort* of someone else, and they were ineffective.

—RECALL an event that you made an *Effort* to attend and were successful.

—RECALL a place that you made an *Effort* to visit and were successful.

—RECALL a time when you successfully moved an object.

—RECALL an instance when you found an object too heavy to move.

—RECALL a time you picked up an object that was lighter than you expected.

—RECALL an instance when you unstuck a door or drawer.

—RECALL a moment when you felt truly successful.

—RECALL a time when you guessed the right answer.

Personal WILL[123] (5.0) is *introduced* from *Self*—along the *ZU-*

---

[123] **Will** or **WILL (5.0)** : in *NexGen Systemology* (from the *Standard Model*)—the spiritual ability at (5.0) of an *Alpha Spirit* (7.0) to apply *intention* as "Cause" from a higher order of reasoning and consideration (6.0) than the thoughts found in *beta-existence*, where it manifests as "effect" below (4.0).

continuum—onto the Physical Universe at (4.0) as its "fullest" 100% *beta-Awareness*. When *Self-directed* its control comes from a higher Source or Cause than *beta-thought*. Any images or thought-forms created by *Self* are *Self-created* purely for personal "consideration," such as to gauge the efforts applied and evaluate results. These are "analytical" processes that do not rely on previous experience or *imprinting* in order to be effectively *Self-directed* from *Self*. In fact, when we are relying on *imprints* and other erroneous data to determine any part of our "considerations," we consider that a *fragmented* state or *system fragmentation*. This can even happen by an individual's own creations—becoming reactive and deluded to the reality of their own thought-forms, literally "losing control" of their own creations. Even a *Seeker* only marginally interested in higher functions of the *Self* would have to admit that this condition seems pretty extreme to continuously go unnoticed—and uncorrected.

After the decision *to be* a *Self* is thought-formed as an *Alpha Control Center* (7.0), the next subsequent level of processed "*Self-Consciousness*" that may be *Self-determined* is the primary "originating thoughts" from *Spirit*, which we refer to as a level of "Alpha Thought" (6.0)—and that includes the highest functions of *Imagination*. Just as there is a *beta-thought* range of frequency determined by the MCC, there is also a higher order of "Alpha Thought" that is closer to "Source"—it is the next highest range of ZU activity able to be *fragmented*—but at a spiritual level. This level of *fragmentation* is also stored—and carried—from *other* interactions of *beta-existence* than only the current "incarnation"[124] or lifetime. This type of *programming* falls under the category of "past-life karma" as carried specifically by the *Spirit* or actual *Self* and is not the same as the "genetic memory" programmed into a *physical organic body*. We are not slighting out the significance of all "past life" *programming*, but it is not an emphasis of our current Grade of *Systemology Processing*.*

---

124 **incarnation** : a present, living or concrete form of some thing or idea.
* It is anticipated that "spiritual fragmentation" will be treated at higher

"Alpha Thought" (6.0) is the level of *Self* making determinant choices and decisions about what is *"to be"*—or else: "what is *to be* WILL'ed *to be*." This is the reason for *considerations* within the high intellectual band of *Games, Systems & Logics*, falling within the domain of Alpha Thought. Only here at (6.0) is an *Alpha-Spirit* making evaluations concerning its *Will* and *Intention* from an *Alpha* source (in the direction of AN)—just as an individual gauges its application of *effort* and *action* to produce *effect* in the Physical Universe (KI). In fact, at each stage of significant communication along the *ZU-line*—particularly the "control centers" where information and energy within each "subordinate system" is *processed*—there is a series of smaller "cause–effect" relays taking place from one system to another. This is one of the numerous reasons our effective application of spiritual philosophy is referred to as "Systemology." The extent of this "direct" *Alpha Spiritual* Cause first turns to *effect* on a spiritual level around (5.0), where it is directed as *Will-intention* to the MCC of *beta-Awareness*.

Will is applied to the MCC (*beta-systems*) as ZU at 100% for that state. When someone is speaking of "Willpower" they are often referring to the determination to apply *effort*, and that is not the same. The *Will* of *Self* is the ability to apply *intention* from a higher order of reasoning than *beta-experience*. Without this, we would essentially be immobile and left to the elements and cosmic forces to decide our fate—and we would be no further along the evolutionary track than a gentle breeze is in creating a turbulent storm. Now it is true that a more "actualized" *Will* is exercising a greater degree of "power" in terms of frequency and influence, but in the Physical Universe of *beta-existence*, "*Will*" is not in "fluctuation" for a *Self-Honest* individual. It is true that the "realization" of personal "Willpower" may be diminished in an individual—but the full vibrancy of frequencies at (5.0) are entirely within the grasp of anyone to tap and express during their lifetime, without even delving into higher intellectual faculties (and Grades) of *Games and Logics* or *Universe Creation*.

Grades of *NexGen Systemology* development.

Alpha Thoughts are equivalent to "Prime Directives" or "Conclusions." They are the most basic points of fact held about *beingness*, and the *"knowingness postulates"* on which all other interpretation of meaning is based. It is then the role of *"beta-thought"* to play these messages out constructively. The premises and support are evaluated only after Alpha Thought has been put forth. The natural ability to construct thought-forms and manipulate images in the realm of thought before acting is an important function of the creative *Spirit*. Without this function, at least as it relates to *beta-existence*, we would exist in states only referred to on ancient cuneiform tablets concerning the *Primordial Chaos* and *Nothingness* of AB.ZU (8.0). It is the *Self*—the individuated *Alpha Spirit* (7.0)—that effectively put *Order* into *Chaos* in the Cosmos,[125] fashioning a "Physical Universe" on which to play out its scenarios.

Personal Fragmentation comes into effect primarily when:

a) an individual is lost in the effects of their own thought-forms (now an *illusion*) and is no longer actualizing *Awareness* as "Cause";

b) an individual is unable to freely change the effects of their Alpha Thought and continues to run on some former programming (now a fixed *postulate* or *belief*);

c) an individual is inadequately analyzing environmental information and evaluating the efforts necessary to change an effect (due to *imprints* and *conditioning*);

d) an individual is unable to charge thoughts with effort to manifest or materialize their Will because emotional energy is wound up (in *imprint stores* and *reactive conditioning*); and/or

e) an individual is restrained from using, applying, or expressing their individual Alpha Thought due to authoritarianism, educational programming, social pressure and other deterrents to *Self-Honest beta-experience* as a result of poorly managed experience.

---

125 **Cosmos** : archaic term for the "physical universe"; implies that chaos was brought into order.

When personal energy communication is clear and *Self-Honest*, the seat of consciousness may be actualized to a higher level, leaving the lower levels to operate on the "auto-pilot" of their respective control centers. This type of "auto-pilot" is actually natural and efficient and there is no concern or danger so long as the pathway is *defragmented* of all erroneous thoughts, programming, conditioning and imprinting. The source of causal control is maintained best from higher "seats" as opposed to descending to fix one's *Awareness* on micro-managing systems that should otherwise be delegated to their respective "control centers"—per their name and function.

The purpose of the "Mind-System" (4.0)—most often referred to as "thought"—as experienced in *"beta-Awareness,"* is to carry out orders to "create" and not to become distracted or lost in illusions independent of *Self-control*. When the control centers operating on the ZU-line are in a condition of *Self-Honesty*, there is no danger of falling victim to one's own illusions. The *Seeker* should not allow themselves to fall victim to their own creations—to be an effect of their own cause. Thought-forms and thought experiments may be used freely to analyze, evaluate and gauge personal efforts—so long as all images and creations formed may be easily *unformed*; anything created must be able to be *un-created*—at *Will*—in order to remain under Alpha control.

PERSONAL MANAGEMENT OF SELF-CONSCIOUSNESS

Self-Actualization is *"unfoldment"* of inherent potential of *Self*. There are many spiritual philosophies and religious ideals that have professed methods to attain this state—yet nearly all of these in the Western world are concerned only with "worldly control" via structured dogmatic beliefs and/or adoption of *a-priori theories*[126] and *postulates* that cannot be

---

126 **a-priori** : from "cause" to "effect"; from a general application to a

"actualized" in this lifetime, and must always be accepted until whatever point they may be validated *after* this lifetime, when whatever truth they profess may or may not be realized as fact. Most intelligent *Seekers* are not willing to wait that long. Therefore, steps must be taken to systematically "realize" these higher states of *Awareness* in the present, even by practice, until we "actualize" an experience of results. *Systemology* can provide the "formula," but it is up to the *Seeker* to plug in various "values" specific to their individual "equation" and then evaluate the results. One or another combination should get through to enact[127] the desired "change." Whether an individual is dissolving their personal *imprints* and *programming* or applying *Self-direction* from *Alpha* to alter universes, the basic pathway and nature of ZU energy is always the same and predictable.

As a result of the decision *to be* in Human form—and by definition adopt the "Human Condition"—the *Self* is adopting the hardware *and* "operating system" conditional to the "Human Condition." The Alpha *processing* capable of the *Spirit* is nearly "infinite," whereas the realization of this higher level by the "Human Condition" is limited by the *programming* that is installed. This *programming* may extend to include the experiential conditioning that is received from the environment and stored as *emotional encoding* or *imprints*, but more often the term "programming" is referring specifically to the *thought-formed beliefs* and personal *agreements, axioms* and *postulates* that are impressed on the "Human Condition" in literal semantics or thoughts. These are the types of statements or Alpha Thought that are later—or in many cases, rather immediately—the foundation for an entire array of associated conditions and agreements concerning Reality.

For example: an individual maintaining programming

---

particular instance; existing in the mind prior to, and independent of experience or observation; validity based on consideration and deduction rather than experience.

127 **enact** : to make happen; to bring into action; to make part of an act.

that they are nothing more than a physical body is then forced to adopt all programming conditions attached to that statement as facts; or an individual believing that all ill effects ever experienced are the result of targeted attacks by an absolutely evil entity will soon find that they are consistently an effect or victim of their environment with little recourse for personal management. This is the kind of "power over us" that personal fragmentation allows when total responsibility as *Self* is displaced with faulty programming.

Once an Alpha thought-form *is* established, it may be reinforced with *Will* as a kind of *"Alpha belief"* (*affirmation, axiom, postulate, &tc.*) charged by the *Spirit* as "intention." This is practically a spiritual equivalent to the forming of *beta-beliefs* and *imprinting*. For example, the term "postulate" is equivalent of an *Alpha Imprint* or supreme "*Post*-for-*You*-*Later*"* to consistently filter a view of Reality through. As a result, all of (*beta*) reality—and interpretation of the same—is then limited to a personal understanding and experience that conforms to that very statement (*&tc.*) in order to maintain the continuity of existence in the Mind. It is not necessarily changing the nature of how things *are* in an objective way, but it is equivalent to how an individual is agreeing to see things and interact with them subjectively—and through that action, often interacting with the nature of the agreements of Reality maintained by other individuals. Unless otherwise *Self-directed* by personal realization, a *Seeker* can only evaluate the effectiveness of their Alpha-caused *Will-intention* when it is directly interacting with Cosmic Law, or the Will of another individual—essentially moments when a *Seeker* may observe:

a) efforts not providing the expected/desired effect; or

b) effects/efforts met with counter-effects from others.

---

* Similarly, *beta-imprints* were described in the "*Tablets of Destiny*" textbook as the equivalency of the RCC displacing *Self* as a reactive mechanism and postulating "I'M *such-and-such*" and hitting "PRINT."

The *Seeker* progresses on the *Pathway to Self-Honesty* by systematically evaluating and altering any inhibition toward actualizing this very progress. In the "Mardukite Path" (*Babylonian paradigm*), this ascent toward Source/Infinity is actualized in an ancient methodology called the "Ladder of Lights" or "Gateways of the Gods"—and it was demonstrated as a *seven-fold* path of cumulative *Self-Actualization*. Although only remnants of the "religious initiation" practices have survived on clay tablets in cuneiform script, we have discerned that each of these "levels" of ascent related to greater personal realizations than simply a series of ceremonial operations. In fact, at each "level" or "Grade" an initiate was expected to have established a higher state of actualized *Self* built upon a former state, then later discard *this* shroud in yet a further development on higher spiritual individuality. "Graded Initiations" found in all traditions of the Ancient Mystery School were later derived, but from which very little of the original significance and actualization remained. The world of esoteric magic, spirituality and mystery *itself* became fragmented and inhibited to factions of highly respected eccentric theoretical philosophers and prestigious dabblers.

Valid *Self-Processing* is intended to repair, reduce or remove any and all significant *programming* or *conditioning* that inhibits a total *realization* that *Self* is the *Alpha Spirit*; that this *Spirit* as "I" is an eternal existence operating a higher-frequency existence than the *genetic body*; and that since the "true" *Self* or *Identity* of "I" is an eternal spiritual existence, it cannot be harmed or hindered by afflictions of a *genetic body*. All programming that is maintained to the contrary to these *realizations* is a *fragmentation* of *Self-Actualization* that leads us away from our ultimate goal of attaining our ideal state.

Although it is true that "meditations" and "thought experiments" are not a true substitution for *actualized* experience, such basic methods of realization actually do assist in effectively "widening the parameters" of potential conscious activity, which allows for greater levels of personal *processing* toward *actualization*. This is why we say that *realization* precedes *actualization*. At each stage or step, we are reaching out, sensing—extending our "feelers"—and then putting our foot forward with strong determination. An experience of positive results allows us even greater certainty to continue on the *Pathway* in the same manner—one foot in front of the other.

REHABILITATION OF SELF-CONSCIOUSNESS

Self-directed focused ATTENTION carries our personal *ZU-energy*—or *Actualized Awareness*—into the cycles and systems of manifestation described throughout "*NexGen Systemology*." These systems—as demonstrated on our *Standard Models* and *Charts*—

> carry "Alpha Thoughts" (6.0) as INTENTION (5.0)
>
> from *"spiritual beingness"* (spiritual consciousness)
>
> into *"mental knowingness"* (mental consciousness)
>
> where it is carried into—and realized for—
>
> the Physical Universe by the *"beta-Thought"* (4.0)
>
> of a *beta-Lifeform* connected on the *ZU-line* (*"Identity"*).

From this degree of *Self-direction* on downward, it is the MCC that is responsible for gauging EFFORT necessary to enact the desired change/manifestation in the Physical Universe—and it is generally "correct" to the degree that it is *defragmented* and *actualized*.

It is not difficult to understand at this juncture, perhaps, the meaning behind the statement that: "best intentions are not enough to gain desired results." Of course, the validation of "Alpha Thought" (6.0) and *Self-direction* of "WILL-Intention" (5.0) are very significant demonstrations of our higher levels of *Beingness* and *Knowingness*. Yet, at the same time, we can easily see how, as this communication signal travels "down" the conduit and solidifies (lowers) in its frequency, it is up to other relay centers to properly cycle and "channel" energies to the extent that these "channels" are free and *clear* of *fragmentation*.

An "*Intention*" is always to *Self-direct* (as "cause") an "*Effort*" to change some condition in existence, such as:—
- to *cause* a thing (or condition) to "be" or "not-be";
- to *cause* a cycle to "begin" or "end";
- to *cause* a system to "start" or "stop";
- to *cause* a "change" in some variable.

This goes as far as to include <u>any</u> *aspect* or "Sphere of Existence" that a *Seeker* has actualized an experience of as "responsibility"—anything that is realized in the domain of personal management, which is the extent (and reach) of an individual's "realized power."

Subjective methods of *Systemology Processing* that incite emotional levels are often useful for rehabilitating *Self* to achieve a clear use of "analytical" *Self-Processing*. These include "*cathartic*" methods and any "*kenostic*" technique targeting or reducing reactive "emotional encoding." However, as a *Seeker* increases in their level of understanding and *Awareness*, greater applications are made to target the "Mind-System" directly, and to make certain no erroneous information is maintained or processed by the "MCC"—the primary *Master Control Center* for experiencing the "interior" of *beta-existence*, or else the Physical Universe. Each step of the way the *Seeker* is actualizing a greater certainty of *Self-Determinism*.

Systemology Processing—whether used alone or assisted—is always designed to increase *Self-directed Awareness*, or else "*Self-Determinism.*" At any point in which *Self* is directing a change, there is an evaluation applied to determine the nature and amount of *Effort* to apply. If "too little" or "too much" *Effort* is applied—or the wrong type—results will not be as desired, but they should be as "expected." Cosmic Law is applied equally to all matters of the Physical Universe.

The greater the *Awareness* range—or field of certainty—

regarding operations of Cosmic Law, the greater degree of *"knowingness"* attained regarding "energetic causality" pertaining to the pure spiritual essence that we call ZU.

A *realization* of this *"knowingness"* may actually be achieved even if an equivalent state of *"Beingness"* demonstrated on the *Models and Charts*—or the *"Circle/Sphere of Influence/Existence"*—is only *Imagined* as a "consciousness projection" *in lieu of* "Total Actualization." A true and full "Total Actualization" of *Self* in Spiritual Existence would then, most likely, no longer require you to control your present physical body—so it is a least something to aspire to with practice.

An individual applies *Effort* in *beta-existence* after a "signal" has been processed down through the ZU-line from (*Alpha*) *Self-directed Will-Intention*. When "true knowledge" or "right knowledge" is processed, then the "right effort" will be applied. It does not matter if it is the act of "turning the page of this book" or to "execute some masterminded creative plan," all *Self-directed* actions require an evaluation of, and application of, *Effort*, followed by an observed "analysis" of the results—preferably without *Awareness* becoming fixed to the *Effects* of results, or expected results.

Actions reflecting personal management (and responsibility) are always directed from *Self*, regardless of where that "seat of consciousness" is actualized from. WILL is always directed from the point of view of the "I" and not the *beta-personality*. If reactive-response mechanisms of the RCC are engaged, a *fragmented* circuit of knowledge will be used to gauge personal *Effort*, but that does not alter the fact that some manner of this energy circuit (*Effort*) is always applied whenever an individual is interacting with the environment—even if inaccurately applied and/or counter-productive.

The decision to use, and the application of, *Systemology Processing* is a *Self-directed* "*Effort.*" This is based on *Will-Intentions* of the *Seeker*—combined with the *Systemology* knowledge base

—to gauge applications of *Effort* to gain the most efficient and effective results. These results are subject to the level of *Awareness* maintained to properly analyze applications of *Effort* and evaluate *Effects*. As successively higher points are *realized*, so then are they *actualized* by the use of "right knowledge" and application of "right effort" to then achieve higher and higher results. Consider an example of applying the *Will-Intention* to push someone on a playground swing-set so that they spin up and completely over the top axis bar and back down.* Obviously if you just were to walk up and push, even a really *really* strong push, you would not yield this result. In fact, experience will demonstrate that "brute force" or "intensive effort" is not necessary or even preferred in gaining the desired effect. However, if we apply "true knowledge" to evaluate our "efforts" and analyze the "effect" properly, we can structure a proper "systematic application" of *Effort* in our mind that will carry out actions necessary to achieve success—in this instance, applying the "right effort" at the "right time" to increase the momentum necessary to haphazardly send a person spinning over the rail. Notice, however, that the original (actual) *Will-Intention* (5.0) from the Alpha Thought (6.0) is entirely unchanged during the whole "analysis-evaluation" sequence of *beta*-events.

Self-determinism is applied to carry out *Processing*, which in turn is actually increasing the available stores of personal certainty toward *Self-determinism*—and potentially higher forms of *Processing*. We have mentioned in passing that *Processes* are operated only until they have resulted in the desired change. This is in observation of the same basic principles of *Effort*, regarding application of any force on environment. A *Seeker* must carry *Processes* "through" in order to have an *Effect*, but no longer. If the same application of *Effort* is carried on past the specific point of desired results, the *Seeker* will not find themselves (ending a "session" or a

---

* This should be considered only as a thought experiment since it is not typically safe to conduct as a physical experiment through to its completion.

specific *Process*) in the highest achievable state. The same is true in other worldly applications too.

It is likely that once an individual *Process* is run beyond its threshold that a *Seeker* will drop back down in states scaled on the *Emotimeter*. Any *Process* can always be persisted longer to earn apex results if *Attention* is not overworked—but once it is overworked an individual will drop in *Awareness*, back to boredom and discontent. Any *Process* or general activity has these threshold points if repeated over and over again. There is the application of *Effort*, a point when the *Efforts* yield a results, and then further application of *Effort*, which feels as difficult and uncomfortable as when originally working toward the desired results. If this happens, switch to a different *Process* before going back to resolve results from the former *Process*. If you get caught up on a "subjective" *Process* involving "recall" or "imagining," change up with an "objective" process of orienting *Self-Awareness* in physical space and in relation to physical objects. And remember:

> *"Objective Processing"*
> assists actualizing states realized in
> *"Subjective Processing."*

One reason is the energetic alternation-shift in focus between introversion and extroversion; another is the cementing of intellectual realizations by associating physical certainties—since we are, for the moment, occupying an *Awareness* within the Physical Universe to the extent that we may be certain in and as *beta-Awareness*.

Most often it is *"Subjective Processing"* that receives most attention and educational focus because of the incredibly diverse dynamics[128] involved. However, mystics and adepts

---

128 **dynamic (systems)** : a principle or fixed system which demonstrates its *'variations'* in activity (or output) only in constant relation to variables or fluctuation of interrelated systems; a standard principle, function, process or system that exhibits *'variations'* and change simultaneously with all connected systems.

throughout the ages were inclined to supplement intensive procedure periods with extroverted ritual and activity—all intended to focus the newly found state of *Awareness* on personal "certainty" to manage the Physical Universe. In many cases today, the "ritualism" is *all* that "New Age initiates" are after, because any higher ideal seems realistically unattainable—often only accepted as a potential byproduct from using an equivalence of "Grade I and Grade II"‡ levels of understanding. Fortunately, for those *Seekers* prepared and ready to receive a higher caliber applied spiritual philosophy than formerly available to *all Humans equally*, the present "Grade III" cycle of "*Mardukite Systemology*" is without substitution.

ASSUMING ALPHA COMMAND OF SELF

Lesson-chapters and *Self-Processing* dispensed in the *present* volume are intended to focus on *defragmenting* personal *encoding* and *programming* developed as your "*beta-personality*"—which is experienced as variations in emotion, thought and the general "attitude" attached to that "personality." There is no reason to fear losing anything here. In fact, you should be able to reclaim quite a bit of your former personal ZU energy "in the process." This is because *beta*-personality is not the "I"—not the *Self* or *Alpha Spirit*. It is a series of heavy garments and fancy dress that each carry a specific package of labels, thought cycles, social roles and programmed expectations.

An individual is "freeing" the *Self* as they systematically strip layers/levels of artificial "clothing" away. These layers are accumulated by the "Me" or "My" as a general "attitude" toward existence—but these are never qualities assumed by the

---

‡ *Grade/Level I—Magick & Physical Understanding, Grade/Level II— Mystical/Religious Esoteric Philosophy & Occult Science, Grade III— Spiritual Wisdom & Mardukite Systemology Pathway to Self-Honesty.* Grade system first described in "*Tablets of Destiny*" (Liber-One).

actual "I." *Alpha Thoughts* of the "I" are best reflected in true statements of *Self*—"I Am" and "I Know" and "I do." But the personality is always a reference to an accumulation of *beta-experiences, encoding* and *programming*—such as those specific to "My Body" and "My Beliefs" and "My Things"—which may contribute to past *imprinting* and present *fragmentation* that inhibits a total expression of future *Intention*.

There are many instances during one's *Lifetime* when "*Self-directed control*" of our ability is "relinquished" to other individuals (or perceived "sources of power"). In fact, one purpose behind the design of basic *Systemology Processing* is to bring *Awareness* back to instances when we have made these agreements of Reality. Then an individual reclaims the *Effort* and *emotional energy* put forth by essentially dissolving it with focused attention at the analytical level—as opposed to a physical or emotional one where it may be more easily restimulated and "hidden from view" by the RCC mechanisms. It is understandable if this material seems complicated to some *Seekers* that have not formerly been taught to process knowledge, energy and experience "systematically"—but with some careful consideration and personal determination, the effects described will speak for themselves, revealing that the effective principles begin to generate results in a relatively short period of time, if carried into "real everyday life" as a personal application and spiritual discipline.

Hindrances to personal development on the *Pathway* are very often times "self-made." They are not actual hindrances at all, but perceived limitations that result from former experience and not from a point of present *Self-Realization*. In fact, much personal *fragmentation* comes from not only what has been done *to* us as a "counter-action" or "counter-measure," but the direct nature of these "counter-moves" themselves, the *facets* and individuals involved, and the level of emotional stoicism[129] and high-frequency integrity a person can main-

---

129 **stoicism** : pertaining to the school of "stoic" philosophy, distinguished by calm mental attitudes, freedom from desire/passion

tain in the face of "*counter-effort*." The reason this so easily becomes aberrated fragmentation for *Self* is because we are, almost immediately, forced to become the *effect* of an external *cause*—to change the nature of our evaluations and intentions due to the "counter-moves" of another person or our external environment. Suddenly, we are taken out of a point of high-frequency causal *Self-determinism* and we have encountered something that *does* have the ability to alter our *Intentions*—which perfectly describes a high-level definition for "*reactive*-versus-*creative*."

It is easy to see how *fragmented* the "Human Condition" actually is in *beta-existence*. Nearly all interactions—when we are not operating in total Self-Actualization as a "spirit"—are sources of *fragmentation*. For example: an individual is going about their tasks, imagining and calculating their work and taking good measure in their walk of life, when suddenly someone does something against them for whatever reasons. This is a problem for anyone not in a position to personally manage *Self* "in the present moment." If a person is to retaliate someone's attacks directly with "an equal but opposite reaction," we are by definition forced into *reactive* measures again, the effect of someone else's cause—and some type of *imprinting* or *reinforcement* may ensue, generally engaging the RCC in the "fight" sequence. "Turn the other cheek" means: don't react; but it doesn't mean stand there and be a punching bag. If we are to do nothing and simply walk away carrying that unused *counter-effort* with us as baggage, it becomes another source of personal dimming—lowering our *Awareness* with another "solid" that we cannot use—and the remains of that energy are now inhibiting the "brightness" of future *Self-expression*. With a little *Self-determination*, this can be resolved. *Self-Mastery* and *Self-Actualization* necessarily includes reclaiming this *bright light*—the *shinning star-light being* that is the real YOU.

---

and essentially any emotional fluctuation.

SUBJECTIVE PROCESSING

Self-directed personal management and high-frequency actualization also includes the ability to properly manage what is generally referred to as "*stress.*" Our society simply takes a stand that "stress is bad"—yet this is only because the "normal standard issue Human Condition" is not instructed on any higher coping skills except to "suck it up" and "hold it in" until the weight of the load crushes you down six feet under. This is no way to operate *beta-existence,* when we consider the balance of "forces" spreading in each direction—AN and KI—holding Reality in suspension. Therefore whenever there is a shift-change of any kind in a state or condition, the simple fact that there is "motion" creates a situation of "stress." This is not necessarily "good" *or* "bad"—"except that *thinking* makes it so."*

> Here is the bottom line—a powerful key to further understanding everything we are demonstrating in *Systemology Processing* and concerning our personal management of existence: directed *Awareness*—any focused concentration of *Awareness*—whether subjectively ("internally") *or* objectively ("externally") oriented...is COMMUNICATION of CONTROL.

Therefore every *Self-directed Alpha Thought* or "command line" from a *Process* is a communication of "control"—so:
>Who is in control of your life?
>Who are you accepting commands from?
>Who is directing your attention?
>Who is demanding that you "listen" to them?
>Who is demanding that you "pay attention"?

These are answers worth recording in your personal notebook, *Seeker's Journal* or *Flight-log book.*‡ Anyone who makes

---

* Quoting Shakespeare—"*Nothing is good or bad except that thinking makes it so.*" (*Hamlet*)
‡ Official *Seeker's Journals* and *Pilot Flight Log-books* specially

demands on us does so with intents of "control." We have stepped into an area of *Self-Realization* that now requires looking outward directly at personal interactions with others as *beta-influences:*

    Sources of *imprint* restimulation;
    Enforcers of Reality agreements;
    Programmer-authorities of information-data; and
    Those who actively use *effort* against us.

Consider also those who advance their energies upon you against your will, or those that reject your own advances and communication.

We are in no way placing "blame" in these exercises—and many times these other persons are "acting" with the "best intentions" in Mind when they operate their lives, to the capacity that they even are *Self-directed*. This disease of "passing blame" carries swiftly between those sharing the Human Condition. At the heights of *NexGen Systemology*, we primarily are interested in identifying systematic causal relationships contributing to *our decisions* to maintain these *Realities*. We aren't interested in "blame." Passing off "blame" leads to a reduction of personal responsibility as "cause" and thus an admission or agreement—from *Self*—to another source as the power of *our cause* and *actions*, which is actually a *re-action* if not fully *Self-directed*.

—IDENTIFY persons that you *presently* consider strong *beta-influences* in your immediate environment (home, school, family, work, *&tc.*) or directly concerning your beliefs and past programming in this present lifetime.
—List them below.

    1.) _____ ‡ ___
    2.) _____ ‡ ___
    3.) _____ ‡ ___

---

purposed for "Systemology: The Pathway to Self-Honesty" (*Grade III-IV*) are future releases anticipated in 2020.

4.) _____ ‡ ___
5.) _____ ‡ ___
6.) _____ ‡ ___
7.) _____ ‡ ___
8.) _____ ‡ ___

—EVALUATE your past/present experiences with these persons sufficiently enough to estimate a basic ZU-line association for each, using *Systemology* education and experience with the *Emotimeter* and *Awareness Scale* as your basis. —*List these "numeric values" for each name, on the corresponding space marked* with a "‡" (popularly used in *Systemology* to indicate either the "ZU-line" itself or "Self-directed Will-Intention on the ZU-line"). *How much and in what ways do you think these persons affect (and have affected) your beta-personality?*

—RECALL the most commonly used verbal statements spoken by each person in your list, or else their most commonly demonstrated emotional state. IDENTIFY any key words, phrases or descriptions you might find when considering these individuals and their influence on you. EVALUATE this data and assign it a value from the ZU-line (which may or may not be the same as what is assigned in the previous list). —*Record this information below for each.*

1.) _____ ‡ ___
2.) _____ ‡ ___
3.) _____ ‡ ___
4.) _____ ‡ ___
5.) _____ ‡ ___
6.) _____ ‡ ___
7.) _____ ‡ ___
8.) _____ ‡ ___

Putting aside your present *Awareness* of these influences for a moment, take some time now to consider your own *beta-personality* and the answers that you gave on your "BAT" evaluation test. Obviously the test itself is very generalized and meant to gauge a numeric evaluation against an objective standard. Therefore, when you answered the questions, you were merely asked about the intensity of specific *Self-determined* values expressed in your lifetime—or the ranking of personal significance in each aspect (category) of *Life Management*. These are *your* scores based on *your* own decisions and choices. Granted, there are many times when *Self-direction* is fragmented by low-levels of *Awareness*, but as the *Seeker* discovers along the *Pathway*, even these conditions were agreements made by *Self* at some point in "Alpha Thought" *after* receiving some type of fragmentation. We later simply go on reinforcing fragmentation with additional agreements and reasoning based upon the same.

After this current measure of *Self-Processing* since your (last) evaluation, it is appropriate to take a thorough review of *Self* and the statements evaluated on the "BAT" in relation to any personal realizations established at this point in your present cycle (run-through) of *Grade-III* work. Consider it a sort of "mid-term exam" if such a concept of school isn't too aberrative for you (and of course, if it is, you might want to "*Process* out" the *facets* from that *fragmentation*). As you perform the following *Analytical Processing*, take notice of any particular *facets* connected with each evaluation, such as emotional reactions, relevant keywords used to describe mental states; also any affirmations of *Self* or "*Self-talk*" routinely made, or even associations often vocalized to others as personal statements (agreements) about ourselves.

> —ANALYZE the present condition of your physical body. —*How do you feel about its condition? What words would you use to describe its condition? What influences your beliefs about it? What influences your intentions in using it? In what ways might it be improved? How might you do*

that? In what ways have you attempted to improve conditions and were not successful. In what ways have you attempted to improve conditions are were successful? What information of it have you observed from others? What words, phrases or expressions do they use? What information of it have others attempted to enforce on you? What words, phrases or expressions do they use? What parts of it are you most certain of? What are your favorite parts? *

—ANALYZE the condition of your physical possessions.

—ANALYZE the emotion that you feel most often.

—ANALYZE whether your typical thought-patterns primarily concern the past, present or future.

—ANALYZE how certain you are in your ability to maintain your existence in this lifetime.

—ANALYZE the responsibilities that you have accepted.

—ANALYZE any responsibilities you are shying away from.

—ANALYZE what additional responsibilities you could accept.

—ANALYZE your certainty in communicating/expressing to others about things you know.

—ANALYZE your degree of trust/certainty about your environment.

—ANALYZE the amount of gossip, hear-say, and falsehoods that regularly appear in your communications/expressions to others.

—ANALYZE your accuracy retaining or passing along information.

—ANALYZE aspects you are presently blaming yourself for.

—ANALYZE aspects you are presently blaming others for.

---

* *Seekers* should consider similar types of questions regarding each line of analysis. Make note of the best or most accurate words and phrases used to describe each part. You should record this information in a *Journal* or *Flight-log*.

—ANALYZE what/who in your environment assists you.

—ANALYZE what/who in your environment threatens you.

—ANALYZE what/who in your environment forces you.

—ANALYZE goals and ideals that motivate your actions.

—ANALYZE your feelings/attitudes regarding others you meet.

—ANALYZE your feelings/attitudes about all Humans.

—ANALYZE your attitudes about all animals and plants.

—ANALYZE your feelings/attitudes about planet Earth.

—ANALYZE your attitudes about the Physical Universe.

—ANALYZE your attitudes about the Spiritual Universe.

—ANALYZE your attitudes about Supreme Infinity.

Compare the data you have collected in this *Self-Analysis* to the previous *evaluations* concerning those considered "influences" in your *Self-direction* as *Self*. The truth is that we carry many goals and ideals that have been imparted to us by others in our external environment—these go on to contribute to our *beta-personality*, which by definition, operates in some degree of *fragmentation* so long as it is retaining significant amounts of erroneous programming. The basic operation of the Human Condition—and our communication with others also operating their own *beta-personality* of the Human Condition—is always a "source" of *fragmentation*, but it does not necessarily have to be accepted as a standard condition or agreed to as Reality. There are energies, efforts and emotions in motion all around us in the Physical Universe—and they may all be effectively managed by *Self* so long as the individual has attained *Actualized Awareness* well above these lower-level frequencies running rampant in our beta interactions.

Self-Evaluation—especially when objectified in comparison to the Standard Model—can be a source of "stress" for some Seekers before they have fully brought into realization the very fact that: none of these attributes of the beta-personality is the "I." And we of course know this, right? We have met its truth in every spiritual and religious and mystical philosophy that demonstrates the most basic tenet of the Human Condition: that we are spiritual beings in some way connected to the function of the genetic body as a living organism. However, the "genetic vehicle" is not the *Self*—and while the entire methodology of *Systemology Processing* and the *Pathway to Self-Honesty* may, from lower-levels, appear to be some extravagant complicated drawn out process, the truth of the matter is: the Seeker is only left "seeking" for as long as it takes to bring this basic fundamental into a total state of realization—and then actualization. Every Process within our current methodology, and virtually every moderately effective developmental or training technique drawn from six millennium of esoterica, all are intended to do nothing more or less than to bring an Initiate to this point. There are a select few that might have access to some type of "magic switch" that could suddenly return them to this higher state —but for most *Seekers*, the *Pathway* is traveled as a gradual process toward *Self-Actualization*. Fortunately, however, the progress does provides enough sure-footed certainty to perpetually encourage a Seeker onward, should they heed the call.

OBJECTIVE PROCESSING

"*Actualization*" of *Self-Determinism* is defined by the level of very literal "*Self-Control*" that is exercised. A person cannot actually exercised true *Self-directed* control in any "sphere of influence" extending beyond *Self*, if the *Self* itself is not managed properly. We may be able to exercise "dominance" or "enforcement" in these other "spheres"—but that results only from intense reinforcement of *emotional* energies, which we have evaluated to occupy lower frequency ranges on the ZU-line.

Emotional energy operates at a range only marginally above vibrations of the Physical Universe (KI)—the level of physical matter and physical energy that composes the very physical organic structure of a *genetic vehicle* or body. This is why *Emotion* is so easily directed and felt between individuals in *beta-existence*—or in other words, "communicated"—almost to the same degree as solid matter. And we have all experienced this to be true in our personal interactions with others. It is also rather easy to detect these fluctuations with even the crudest biofeedback technology of our current age—a fact that only increases an individual's certainty of *Subjective Processing* if employed in conjunction with it. This is not absolutely necessary for success, but it is an option for *Seeker's* looking to find external technologies that directly validate evaluations of progress on the *Pathway*.

Emotional energy is closely tied to the efforts exerted as actions in *beta-existence*. There is nothing inherently good or bad about it—it is a necessary condition for expressing higher level thought into lower level physical activity, and that is all it is. What we are concerned about in *Processing*, is the degree of personal "control" that is maintained when managing physical activity. This is why it is so important to understand and *defragment* the RCC at this present step on the *Pathway*. Because the energy channel is a two-way conduit—meaning information is sent back to *Self* along the same routes once it receives "results" of its efforts (or the efforts of others) to analyze. This information coming in (or input) from the environment (or consequences of action) necessarily passes through our *emotional range* before again reaching *analytical Awareness*. And this fact cannot be overstated to the *Seeker* until it is realized. A close examination of all esoteric texts and practices regarding spirituality or mysticism will reveal unilaterally: "Self-Control of Emotion" is perhaps one of the most fundamental steps on *any* version of the *Pathway*. This is achieved incrementally when working with "SP" *Processes* until it is mastered. Let's try a little practice.

> —IDENTIFY an *object* in your environment that is neutral (harmless) and of which you are primarily

indifferent to. [Start with something you have absolutely no attachment to, such as a rock or paperweight. Teachers operating within the "Ancient Mystery School" often used something like a candle or a pebble.]

—LOOK at the *object* with your full attention and *Awareness* and ANALYZE the extent of your indifference and neutrality regarding its condition "*to be.*"

—IMAGINE that the *object* is suddenly the most awesome, useful, positive, valuable "thing" presently assisting your existence. Spend several minutes *realizing* this until your emotional frequency raises up to the highest "elated" degrees of joy and enthusiasm that you can possibly actualize. ANALYZE any additional *facets* directly experienced as a result. [This is *your present* (4.0) on the *Emotimeter.*]

—IMAGINE that the *object* is now suddenly an even more amazing, beautiful, breathtaking, intricate and perfect demonstration of manifestation. Spend several minutes *realizing* this until your elated state has increased into appreciating the *object* as the most archetypal piece of art, music and verse all combined in one. ANALYZE any additional *facets* directly experienced as a result. [This is reflective of *your present* (5.0) on the Total Awareness Scale, which is in the lowest *Alpha sphere*—of Artistic Expression, Creativity and Aesthetic Appreciation—but still milestones above where people are led to evolve in societies built around "material economies" only.]

—IMAGINE that the *object* is suddenly the most trivial, useless, negative, distracting "thing" presently attacking your existence. Spend several minutes *realizing* this until your emotional frequency lowers down back into the range of *Emotion*. The more intensely this *Awareness* is applied, the lower the *degree* that is actualized, including a pass-by of the former feelings of indifference, into feelings of anger, rage, perhaps also even fear if a true *realization* is reached that this *object* is suddenly pure anathema to your existence, to which point we will eventually be able to be so low that we are complacently "at one" with the *object*, and are subdued by its control. ANALYZE any additional *facets* directly experienced as a result. [This could potentially demonstrate your present "full curve" or "*Pitfall*" potential from (2.5) down to (0.1) on the *Emotimeter*—although the step is often switched before realized to its entirety.]

PRACTICE shifting your *attitudes* and *emotions* between states as well, making sure before the end of the session to leave yourself in a high-frequency state—though generally indifferent, again, to the *object* itself. The *object* is a focal tool only. YOU can be in control of these states of *Awareness* at will—at any and all times. (And *that* is the goal.) Make certain to end the session at high-frequency *Awareness* after practicing emotional fluctuation.

Although quite basic and innate, the practiced *self-controlled* ability to shift *emotions* and other *states* at *Will* by *Intention* is something not realized equally by all. It takes practice—as do many of the methods and experiments suggested in *Systemology*—and the logic behind the structure or systematic design of these *Processes* is not always clear at the start, or even always readily apparent at this *Grade* of educational materials—but the bottom line is that they yield progressively effective results toward our ultimate goal, and it is for that reason alone that they *are* suggested. Otherwise, the *Seeker* is left with the same intellectually described verbiage as recorded over thousands of years, but which leaves very few true *Self-Honest* avenues to gain accurate realizations. This type of "education" and "practice" has not been readily available or present among the masses for thousands of years—and generally each time that it has resurfaced, it has been quickly blotted out or corrupted, or most often reduced to a series of proverbs, morals and dogmas that no longer carry the full extent of the original message.

FURTHER APPLICATIONS: The previous exercise may be applied several times, as cycles *on the same object* and it may be applied to a different *object* during the same session. You may then even use the *two objects* at the same time, choosing one to be the "positive" and one as the "negative" and then alternating the focused attention of your beliefs, feelings, attitudes and intentions about each as rapidly (but fully) as you can. To extend this practice further, one would simply select *two of the exact same object* and apply this alternation, but with arbitrary selectivity of one from the other, named uniquely from *Self*, or as "*Paperweight-A*" and "*Paperweight-B*."

Although we are using this method of *Objective Processing* to draw the *Seeker's* attention toward *Self-Control* of *Emotional* energy—and its notorious slides and swings—the same technique may be applied to lessons illustrating "thought association" (or "associative thought"). This is the basic scheme by which we "separate" things in our Mind-System as being: this *or* that; this *and* that; this *but not* that; and so forth. These "associations" from thought are also linked to our "emotional-effort" systems—they interact with one another and the total result is equally within the domain of total *Self-Control* so long as the *Seeker* is even *Aware* of the "thought patterns" and "emotional cycles" that reoccur. While it is true that all *things* are not of equal importance, the power that is wound up in "associative programming" that we once agreed to—though perhaps forgotten about and pushed off the screen of present everyday surface thinking—is still "running in the background" of our daily *beta-existence*, of which our "basic" level of understanding will still consistently return to (as an equilibrium). This is one reason why *defragmentation processing* is so significant in all mystical and spiritual schools that have ever contributed to our modern day library of esoteric material.

SELF-DISCOVERY OF HUMAN ABILITY
[EXTENDED COURSE]

This is a transcript of a lecture given by Joshua Free on the evening of October 31, 2019. Content of this lecture introduces course material retained by the NexGen Systemological Society (NSS) from the International School of Systemology (ISS). It is included within this volume to supplement previous chapter-lessons and for the benefit of Seekers (and Pilots) interested in experimenting with practical applications of Mardukite Zuism and Systemology (ZU)-Tech alluded to throughout this volume. Additional textbooks, guides and manuals are currently in development to advance further work within this applied spiritual philosophy of Systemology.

△ △ △ △ △ △ △

"Human perception involves *coding* even more than crude *sensing*—as *Thought* is *abstraction*. Language, Mathematics, the schools of art, or any system of Human *abstraction*, gives to our *mental constructs* the structure, not of the original fact, but of the *symbol-system* into which it is coded... The basic symbols of *magic*, *mythology* and *religion* are so simple that only the pernicious habit of looking at 'profundities' and 'mysteries' prevents people from automatically understanding them almost without thinking..."
—Robert Anton Wilson, an esoteric philosopher, author of "Cosmic Trigger" & "Prometheus Rising"

When I first began giving mini-workshops at the Mardukite Offices in 2009 regarding what is now called *NexGen Systemology*, I would usually begin with an exercise that I, myself, was given when I was taking some creative psychology classes in the 1990's. Taking a clear clean piece of paper, you would write the words "I AM" in big capital letters in the middle at the very top, and then you were asked to basically fill the page with all of the associations you might have with that—or else all of the applicable ways you could complete the statement as a sentence without inventing something ridiculous. In fact, you might just go ahead and try that. Because it is funny—you know—people really, *really*, want to fill out that page. People *really* want to attach things to their *Identity*. How else can we gauge ourselves against our fellow man, right? How else could we possibly orientate ourselves in a domestic situation, or in a group or as a participant on the planet without a series of titles and roles, right? That's a *"personality."*

The "Self-Discovery" lessons and *processes* are intended to do that very thing: "discover *Self*," as in "uncover *Self*." We are peeling away garments that have been added as *beta-personality*—and I can assure you that the "I" that is *Self* is never lesser for this. It is quite possible at this point of a *processing* cycle that the *Seeker* has at least gained a partial evolution of consciousness or greater realizations about the difference between the *corporeal body* present in this *beta-existence* and the *Alpha Spirit* that is the "I" or *"Self"*—that some call the "Higher Self." But that term bothers me as an application to *Systemology*, because we are only recognizing *one "Self*," which is the "I" at (7.0) and of which we may project an *Awareness* of to any "degree" along the ZU-line, but *Self* remains. In fact, once you systematically peel back all *associations* to "I AM" put on that piece of paper, only the "I AM" (7.0) will remain.

It would be an easy exercise to say—from an intellectual level—"okay, now I know the trick: I write 'I AM' and I just leave the page blank and, hey, look at that, I've just achieved enlightenment. I'm Self-Honest. Um. Where's my fancy certificate?" Okay. So, if you could do that, and *actualize* it, you're *there*, man. Good for you. But to most *Seekers* working their way up the *Pathway to Self-Honesty*, such an effort is first *realized*—and that *realization* is incrementally reinforced before it *actualizes*. Yes, the *Knowing* is an important step, but it is not the final step. You still have to *Be* it. The goal, would be, actually, as being in and as *Self*—the *Alpha Spirit* at (7.0)—you *are* actually achieving that state of "I AM" in a flash moment, because that is exactly what the condition of that state *is*, and the only way in which we can describe (7.0). It is the I-AM-SELF, pure, clear and *defragmented* of all fluctuation and condensed energy. It is only immediately below that, in the band of (6.0) to (6.9) that we have Alpha Thought connected to *Games* and *Universe Systems Logics*—all that good stuff which provides anything of consequence to attach to I-AM-MY-SELF. This means that everything we are attaching from this level on down is attributes of "*Me*" or "*My*"—but not "I." Below this, of course, we find that "I am Myself and there are others"—there are other "Not-Me's"—we have others to play our *Games* with, and so we exert our *Will-Intention* at (5.0) to enact our moves in *beta-existence*, which as we know, is directly experienced as degrees of *Effect* below (4.0).

Our *models, scales, ZU-lines, charts*—everything I have worked to careful synthesize from workable models alluded to on the most ancient tablet writings from the inception of *this* modern conception of the Human Condition and its systematized living conditions—all of this is to provide a relevant model to orient or direct your *Awareness*, for which you have not been impressed with by any other tradition or system. These *models* and *charts*, however, will only serve an individual to the extent that they are *realized* into *actuality*—otherwise we are simply left with remnants of a very ancient *Kabbalah*, one that even predates the more famous Semitic version, which resulted in countless spiritual and magical and religious interpretations of lore from the Ancient Mystery School. If these were to suffice—if the manner by which the "true knowledge" and map for the *Pathway* had already been so clearly given—we would find little value in establishing *Nex-Gen Systemology* for the 21st century. But, that is not what we have found—either at the office or among those who have put this work into practice. We have found more workable and effective results with sweeping progression unlike anything else that has been delivered so accessibly to humanity. And believe me, I have looked all over to be sure.

Anyone who is familiar with my former work, which we are now classifying as *"Grade I"* and *"Grade II"* materials, knows fully that I am no stranger to all of these diverse avenues from which traces of this *New Thought* may be found—but they were so badly fragmented as exclusionary methods, that the whole of it had to be discarded to achieve any solid progress in *"Grade III"*—which is why it took eight years to bring this work to the public from the underground. It could not be found workable with any direct semantics or system ties to practices developed after fragmentation of the Ancient Mystery Tradition—and even then, only sparsely discernible from the very few surviving cuneiform artifacts found to be actually valuable for our purposes. However, it has been pointed out to me on more than few occasions that what we *did* accomplish at the *"Grade II"* stage of "Mardukite Chamberlains" work *was* a complete reconstruction of the ancient BAB.ILI Star-Gate System or else the original *Kabbalah* of Babylon, which *is* very literally and directly the predecessor and inspiration for the more popularly known *Semitic Kabbalah*—and we have seen that there is at least *some* valid effectiveness in *that* version, we should expect *no less* from a more direct purified clear version, right? So, the semantics, framework and cosmological model of the *Babylonian Star-Gates* became the basis for the *Standard Model* of *Systemology*—and there is little doubt, as you have seen, that this works. And the *realizations* and *Processes* drawn from the same are yielding *actual actualizations!* This is, at the very least, exciting for me, because I have been out to find these results for, literally, a quarter-of-a-century—since 1995.

△ △ △ △ △ △

I have spent some considerable time systematizing the "true knowledge" that has come down to us through the ages—and the further we went, applying the logic of *Systemology* all the way, the clearer this map became until we reached *Infinity*. Tremendous *Efforts*—more than I would even care to admit—went in to making certain that this *Systemology* was complete and perfected to the extent we are capable of delivering it as a "core." As a *Grade*, the material must stand for itself as a "core." Some of you are familiar with the previous *Grade's* "Mardukite Core"—which is a name that we have attributed to *that* "stand-alone" collected body of *Mesopotamian*-oriented material for *Grade II*. When we compare the magnitude of the "Mardukite Core" and the previous cycles of work concerning Western Magick and Druidism—as reflected in the *Grade I* materials—a *Seeker* should expect that *Grade III*, the work encompassing "*Mardukite Systemology*"—officially called such in honor of its roots from "*Marduk's Tablet of Destiny*"*—to be an evolutionary milestone of development on this path I set out on twenty-five years ago.‡

The next *chart* that I want to supply you with is more intuitive than the others because it is meant to be, in effect, "superimposed" transparently onto *Standard Models* and the *ZU-line* and so forth depending on the perspective being realized. I have referred to the concept many times before—and this is certainly not the only version of such lore that you can find if you go digging—but it applies quite adequately to our purposes for the *Pathway to Self-Honesty* and as a direct bridge to *Total Self-Actualization*, which is probably the goal for at least a few of you.

[*Audience Voice: "So there is a difference?"*]

Between "*Self-Honesty*" and "*Total Self-Actualization*"?

[*Audience Voice: "Yeah."*]

---

* *Tablets of Destiny: Using Ancient Wisdom to Unlock Human Potential.*
‡ Officially begun in 1995, when the present author—not yet even in high school—received his original initiation to Douglas Monroe's "Pheryllt Druid Tradition," attended Deepak Chopra's "Way of the Wizard" course and then subsequently released his first underground publication—*The Draconomicon*, while writing under the pseudonym of "Merlyn Stone" at the age of 12. Joshua Free's entire legacy is thoroughly documented and has received worldwide underground recognition for over two decades.

Oh, sure! Yes, definitely. Our aim is always *Total Self-Actualization*, even if only "imagined" as a means to greater "realizations." This is accomplished, in our methodology, by systematically *defragmenting* the conduit or *Pathway* that literally "leads to" *Self-Honesty*. In view of the fact that we are most concerned with what is right in front of us—being at this time: The Physical Universe—it stands to reason that we have to deal with these physical and emotional and yeah, all the way up through the *Processing Center* of the range we carry as "*beta-Thought*," and by that, for those of you that might be confused, we aren't concerned with the scientific semantics of "brain-waves," but concerning the contrast of the *Alpha* and *Beta* states of existence. By referring to them as such, we can later apply a more refined use of general semantics such as "physical" or "internal" or "thoughts" or what have you, but we describe these in most basic terms as *Alpha* and *Beta*. We can then make associations concerning, for example, the *direction* of AN and the inert entropy of KI, but we do this in relation to *Alpha* and *Beta* universal existential conditions and not purely in concepts like "heaven" and "Earth" or "up" and "down" or "physical body" versus some kind of "mental body" and some other "body." People are really attached to *these* ideas about bodies, you know.

The concept we have put forth—or originally put forth back in 2008, I think it was—as "*Self-Honesty*" is like a midway point of "*Total Self-Actualization.*" To be honest, and I have said this before, I do not know that your vibrations could be sustained as a *beta-existence* when you hit that point as an Absolute. But see, what happens here, is energy operates as a constant in abundance. Switches are "on" or "off" and then, well, then you have resistors and loads and dirty wires and such. I'm only using electricity as an example. The point is: the great teachers speak of abundance and affluence and we don't even need to start mathematically calculating friction or counter-efforts from others in order to realize that whatever results we want to achieve, our mark needs to be at least just beyond that. Even with seeing our way clear through *beta-ex-*

istence at (4.0) we still speak of *Will-Intention* and the application of "higher aesthetic" levels of realization at (5.0).

For purposes of *Systemology*, we marked the point of "Self-Honesty" as (4.0)—the same as the MCC and the contact threshold point of ZU between *Alpha* and *Beta*—or typically what we refer to as the distinction between the Spiritual (AN) and the Physical (KI), which generally applies. It is assumed that if you can more or less clear out the debris of *beta-programming* then you are at least up to a point of a *Self-directed Will* making contact with the *genetic vehicle* you are operating. Even if we take the *body* out of the equation—taking the entire idea of a physical existence or *beta-existence* out of the equation—we are still left with *Self*. The *Identity* at (5.0) through (7.0) seems to exist and operate as a individuated *Alpha-Spirit form* (7.0) with *intelligence* (6.0) and *will* (5.0) completely independent of any physical form and perhaps even without a physical universe to interact in. So, it is a valid question, because this is all significant for incorporating this chart—which is really a two-fold chart, and that is why I say it requires some degree of intuitive thought to see past the labels only and consider the applications in *Processing*—and possibly even the relevance of this education at higher levels of *actualization*.

I actually want to share another quote here—had it in mind, put aside here—as I was waiting for the right time or most relevant time to slip it in for some addition food-for-thought. But, here we have Carl Jung, quite an ingenious and creative fella when you consider he came out of German psychology—but yeah, so he wrote this quite intriguing autobiography of sorts titled "Memories, Dreams and Reflections," and one of the interesting passages there actually appears in its Prologue, where he says:—

> "My life is a story of the self-realization of the unconscious. Everything in the unconscious seeks outward manifestation, and the personality too desires to evolve

out its unconscious conditions and to experience itself as while. I cannot employ the language of science to trace this process of growth in myself, for I cannot experience myself as a scientific problem."

This new graphic chart was originally unlabeled and only contained symbols and numeric values. I've modified it here for examples—but its essentially the same whether used objectively regarding the *"Circles of Existence"* based on our perceptions of levels of manifestation, or most definitely applicable to our concept of *"Spheres of Influence"* or else *"Spheres of Interaction."* In either case, the divisions work their way out from the center as *Self* perceiving "I" from the *beta-perspective* working its way through the energies between our *actualization* of the Identity as *"Self"* and all other realizations or levels of *Life* that exist between where you now sit and *Infinity*. As a *Model*, these *"Circles"* or *"Spheres"* also perfectly demonstrate the harmonic flow of interconnected energies plotted on a graph using both positive and negative on the axis lines. The *maths* behind all of this are always there for those who work that out, but this knowledge does not in any way increase the effectiveness of whether or not the car goes when you push the pedal, so long as all the other necessary conditions have been met.

There is a lot of background engineering to *Systemology* that we will get into in later work. Right now we are focusing specifically on what is *most critical* to raise consciousness activity levels in the direction of *Self-Honesty* and higher level *actualization*. Not to mention—I should point out here—that the "mysteriousness" of the mechanics dissolves the more you work with all this. That's what is making our progress so exponential now—is because the more our researchers worked with various materials released earlier, or in the underground, the clearer the steps ahead became. Some steps have not been actualized and await higher graded work, but they have been *realized* and the *maths* and *logics* of their harmonic demonstration on the *Standard Model* is valid. So, we at the

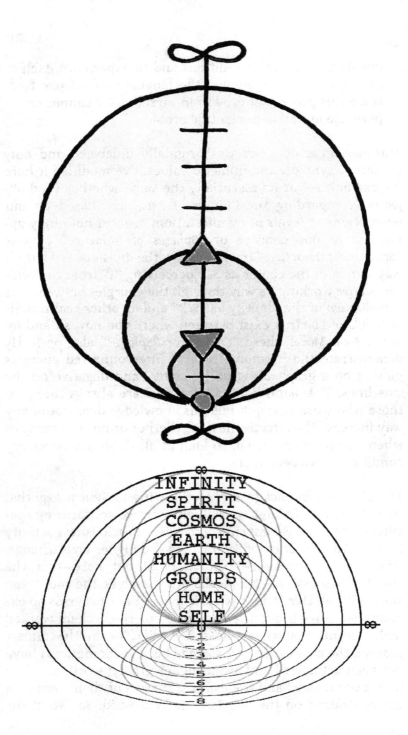

very least *know* with certainty that we have something *very real* to look forward to.

The purpose of introducing the *"Circles & Spheres"* is primarily this: the *practice* of *Beingness*. That is to say: the *realization* of *Beingness*. We've touched on the physics and metaphysics of universes for the *Standard Model* and the integration of consciousness and *Life* as the ZU-line to provide coherent symbols to chart or map from here to *Infinity*. This is one of the reasons why some mystical or magical schools and philosophies have placed so much emphasis on cosmological models and *kabbalahs*.

We established the prehistoric Mesopotamian *Kabbalah* in the "Mardukite Core"—and we codified a highly systematized interpretation of the same in our *Systemology*—but I'm not here just to give you another *Kabbalah*, all nicely decorated and, you know, with calligraphy and such. They look *real* nice in books and they are great for systematizing the occult, because well, they have all of these fancy "Divine Names" and words to utter over there—and they assign them to all these different positions. This is all in *Grade I*, mind you, so some of you know what I mean when we talk about "magic" or the "New Age," yet a few of you now here are raising an eye-brow because *Systemology* isn't about the kind of rituals and seances you've seen on T.V.—and we aren't taking it there; it isn't about that. But, if you are so inclined, *Systemology* qualifies, explains and demonstrates what Humans call "magic" more accurately than the *"magic books."* That wasn't—that's not my goal, because, its funny, it helps clear up things about "metaphysics" and "spirituality" and "religion" and "psychology" and the "Human Condition" and—well, everything! The fact that it validates something that you are into is great, but I'm delivering *Systemology* with a greater purpose than that—whatever *that* is. I'm not here to grant you another *Kabbalah* —I'm here to grant you *Beingness*.

△ △ △ △ △ △

If we look at this chart of *Spheres* and *Circles*—see, originally I wasn't going to even put labels on it—it is the *Standard Model* of universes and the personal *ZU-line* that is already given in our material, but which is now examined for "*interconnectedness.*" Everything *is* connected together as a continuity in KI or an infinity of AN—but between here and there, we *are* individual *Spirits*. I'm not particularly attached to a physical body, but I do like the idea of being an "*individual*" Self-Actualized spiritual entity, up here at (7.0)—at "I AM." That's a pretty good place to be. You aren't bound to anything specific—you aren't exactly blurred into the oceans of consciousness like a water-drop in the Infinite Abyss of Nothingness, but you can probably get a *sense* of it from there. Its rather difficult to maintain an *Awareness* for very long up here and then still be tinkering around at lower-levels of the *genetic body* at the same time. But, guess what?—Using these keys alongside *defragmentation* allows for the *realization* of Beingness—while still hanging around that physical body you got going on there. Because, you know, these bodies have some use for *playing games*, and that's more or less all we are here to do in *beta-existence*. Just don't forget that you *are* "playing" a "*Game*," because we often forget that point when we enter a lifetime and then get very caught up in this other "My-body-my-self" *Game*.

SYSTEMOLOGY PROCESS "SP-2B-8"[*]
(DESCRIBED)

If you are familiar with the *Standard Model*, or the *ZU-line* in "*Mardukite Zuism*"—or a clear understanding of the "seven-plus-one model" derived from ancient Babylonian cuneiform tablet literature—than the "*Sphere and Circles Models*" are really more of a visual aid to what you know. But, it is actually more than this. Because what we are wanting to *realize* from this is our ability to *Be*, to project *Awareness* as a *realization* of every point of potential existence between here

---

[*] A designation later assigned to the process described in this lecture.

and *Infinity*. The more we can *realize* our interconnection and influence to these other *Spheres*, the more we gain a certainty of our *Self-directed* ability to *Be* the *Cause*—to literally *"Be Your Own Reason"* to the very limits you are able to conceptualize that, which for most of us, would be next to *Infinity*, once you get that far. This is not some arbitrary exercise—or some fanciful imaginative play without purpose. This also prepares the *Seeker* to start thinking in even higher realms of Alpha Thought related to *Games* and *Universe Logics*. All you really have to do here—and the *Pathway* map lets you off easy on this one—is essentially: *"fake it 'til you make it."*

So we have the eight *Spheres* and *Circles*—represented in their *Absolute Totality* as positive and negative values for all *space-time energy-matter*. On the subjective scale of *Self*—where we are operating *beta-existence* from the *Awareness* of *Self* and *Self-determinism* as "Cause"—we place *Self* at the center of the *model*. This is equivalent to *Self-Mastery* of the level of *Self* actualized in *beta-existence*—meaning: to the extent that we are able to actualize *Self* in the Physical Universe (KI). This goes back to what we were saying before about the difference between *Self-Honesty* actualized in *beta-existence* and the idea of *Total Self-Actualization*, which is a state actualized exterior to the Physical Universe and "Physical Body," but of which, for example, *Processing* using the data of the *Spheres* and *Circles* as a base, may at least allow *realizations* to the extent of our abilities while in *Beta*. These same *spheres* or zones relate, in part to the *Systemology* we have described, and in part to the old twentieth century New Thought which started tackling these issues over a century ago. But we have come a long way in our understanding since then—or at least our accessible potential for understanding. It must necessarily be that way, because *systems theory* dictates that the complexities of worldly systems will always extend just one point beyond the understanding actualized by its creators—and we are in some pretty deep convoluted bio-waste at this time in human history.

This *chart* then places *Self-existence* in the center at ("1") overlapping the "KI" continuity line of the Physical Universe which is "zero." And you will see that the *spheres* include both a positive and negative aspect to balance the equation. The sub-zero ZU-line is dealt with more critically when you are concerned with the creation of *universes*, the *maths* of *logic* and *games*, and the fabrication of simulacrum[130] or *beta-bodies* to experience *beta-existence*. Well, right now, you're already in a Physical Universe; its got some *maths*; you're playing a *Game* —probably a few; and you got that *Identity* all anchored up to the control of this body here. We might as well deal with the conditions we have here on our plate first, and then we can dissolve some of that *fragmentation* with the same steps and learning that earns us higher positions of *Self-Actualization* on the *Pathway*.

The higher the frequencies you truly *realize*—with more than just the intellect and "mental consciousness"—the easier it is to simply dissolve the lower-range energies that are inhibiting and *fragmenting* expression of personal management. This is one reason why a good *Processing* regimen is going to alternate between resurfacing and managing an individual's past and present concerns and then focusing on some higher spiritual lessons and exercises to allow for a greater state of *Beingness*—and not simply the *Knowingness*.

I'm sure some of you have spent more of your life dedicated to scoring over dusty old books and stacks upon stacks of "magical correspondence" lists and charts then on anything else. It gives a certain boost to one's individuality, for sure. It is all quite interesting and even effectively demonstrates many basic valid premises concerning the *Pathway* we are on now—so that we may selectively borrow from it at *Will*. But!—

---

130 **simulacrum** : an tangible image, facsimile or superficial representation that carries a likeness or similarity to someone or something else; in *NexGen Systemology*—the *genetic vehicle* or physical body is an example of a "simulacrum" of the true *Alpha-Spirit* or *Self* (I-AM), which otherwise has no tangible form in *beta-existence*.

walking backwards nine times around some gravestone while chanting "hockety pockety" between coughs on jasmine incense smoke—yeah, it's a lot of fun and it *is* a considerable step in the *Grade I* materials when an *Initiate* has first begun to *realize* that there is *something* beyond the Physical World. But that is *all* that you can really glean from that work if you are applying *Self-Honesty*. All of the rest of it—you know, the *really valid stuff*—well, we can do those exercises now here in *Systemology* without concerns of getting entrapped into some paradigm of "mumbo-jumbo" or adherence to some distant greatly misunderstood mythological pantheon of *gods* and so forth. The only *spirit* we want to "conjure" is called *Self*. Otherwise, that's not what we are about in *Systemology*. In *Systemology*, we can choose to validate or *obliterate* such mystery *at Will*.

Make no mistakes here about the *Self* at ("1"), which I should add, also includes ("–1"). This is not directly the same as the emotional level of (1.0) on the *ZU-line*, but rather in relation to (1.0) as vibrations of physical energy-matter in action. Therefore, we are concerned with the existential *Self* as a physical being and not the "equivalence of (1.0) on the ZU-line. Okay, now, *Self* seems pretty critical in these equations because if you take that out of this *chart*, then you don't have a *chart*. So, consider *Self* at ("1") as *Awareness* of *Self* and efforts of continued existence *for-and-as* "*Self*" in the Physical Universe. To clarify one last time: This *Self* at ("1") could be actualizing any degree of ZU-energy, like (2.2) or (3.5), whatever, and it will still be right *here* on this chart at ("1") so long as it exists.

The greater we *have* actualized *Self*, the greater our "reach"—and *that* is what we mean by "Spheres of Influence"—or "Circles" if you prefer the two-dimensional approach. So, that means that being statically placed here at ("1"), *Self* remains in place there and all we are doing is extending the *reach* of our *Awareness*, not the displacement of *Self*. This is where I lost some people in former workshops, so I am going to explain it this way: think about what you have seen charted as the *Alpha Spirit Self* at (7.0) in former models. Now imagine that instead of moving up into *Infinity*, you are directing that whole scale of personal existence into an alternate reality in the position of ("1"). Your *Self* is still (7.0) in *Alpha* state—the individuated I-AM of the *Spiritual Universe*. It is now projecting its *Awareness* into a *beta-existence* of a specific physical "dimension" in position ("1"). It may raise its level of consciousness in regards to this new *playing field*, but whatever happens to it, whatever it decided to do with direction of *Awareness*, the true "core" or "Eternal Self"—back on this other *Standard Model*—remains at (7.0) unchanged. Therefore, the *Spheres* and *Circles* are specifically related to our *experience* as *Self* in *beta-existence*.

The entire *Identity* then—the entirety of a personal ZU-line as *Self*—is manifested in a Physical Universe at ("1"). What it represents is the individual's personal ability of actualized *Self-Realization* and *Self-Control* in terms of personal management, using the total *Self* to form the *"causes"* of any and all *Being*. This isn't fancy word play—we are very seriously dealing with "control" from the point of view of a very powerful force called the *Alpha Spirit*; a *force* that can very effectively direct its *Will-Intention* onto existence. Although many of us have passed off *control* of this *force* to some other *direction*, but this entire *Systemology*—and your personal progression on the *Pathway to Self-Honesty* is dependent on the quality of your own *Self-control*.

When we are talking about *direction* and *Will* and *intention* and —well, we even have all these "*control centers*" in our vocabulary now—it should be very clearly evident that the baseline fundamental behind all of this work is CONTROL and how CONTROL is operated and *who* is doing it. You could probably reduce a definition of the whole field of *Systemology* to this without even using the word "system." When you—or when you *Pilot* your *Seeker* to—direct attention to things, or select things to give attention to, and change your attitudes about things, *all at Will*, mind you; this is all boiling down to "control." Using these methods to control and direct—and allowing you to choose how to control and direct your attention and actions—this is all going to increase a *Seeker's* certainty to *control* personal management of *Self*, and then of course the extent of our *reach* on *control*.

It's not hard to get a realization on *Self*—because, well, *there* you are. So, then we move into the next sphere—and here we create or manifest and maintain some type of *shell* for our *shell*—and we call this ("2") the domain of *Home, Family*—basically our domestic security when we flashback to what ol' *Maslow* was trying to say with his "Pyramid of Self-Actualization." Once we realize that we are *Self* "in" or "enshrouding" or "controlling" this body in *beta-existence*, our field of influence immediately manifests toward a condition best suited for perpetuating existence—specifically the ongoing personal continuation of *Self*—and that requires a *Family* or a "domestic" situation of sorts.

Just as *Awareness* of a *beta-Self* "identifies" with a *genetic body* for this *beta-existence*, so too does one easily take on the *Identity* of "family" at ("2") as another *fragmented* level enshrouding *Self*. But, if we are maintaining full *Self-Determinism* at ("2")—the same analytical efficiency we are achieving for *Self* at ("1")—without *losing* control of our *Awareness* into the second sphere, than it may be used as an effective survival mechanism. One reason I did not drop this *chart-model* on the material earlier is because it necessarily requires a *Seeker* to understand the dynamics of, and maintain a full realization of, all the intricacies regarding the *fragmentation* of the *Identity* of *Self* at ("1"), before treating the next closest "Sphere of Influence" as its own "*Identity*" as a "consciousness"—complete with its own associative degrees of *Awareness* and *beta-fragmentation; Imprints; encoding;* all of that.

This idea of *Identifying* with anything other than *Self* from its highest point of independent spiritual *Self-directed* existence is a source of *fragmentation*. Just as you can over-*Identify* and "lose yourself"—as they used to say—at ("2") in the *Family-Home-Domestic* band of influence, well, it obviously doesn't stop there. We move along this *chart* of *existences* rippling out from this epicenter of *Self*, and after *Home* we have the entire domain of *Organizations, Societies* and any "identification with a group" out here at ("3"). For any given individual, this is probably composed of a series of various "rings" or "social circles" that may be independent of one another and still overlap in various ways. These are all indicators of a "group mentality" as an *Identity*, which naturally carries its own *Awareness* level—even its own numeric value from the *Emotimeter* and so forth. Basically any way in which we might treat *Self* as an "individual" on the *Pathway to Self-Honesty*, we can equally evaluate one or another of these *Circles* and *Spheres* as an entity. There is no question about that. And it makes for an excellent *Systemology* exercise.

The remaining "Spheres of Influence" correspond to greater and greater circles of inclusion. For example, our *Home* is a basis for *Self* in terms of a "base of operations" for the Physical Universe (KI). It includes others who are in our vicinity or proximity—those within our immediate sphere of close personal influence. This a two-way interaction of influence and control—and it is greatly *fragmented* because most of the *emotional encoding* and *Imprinting* that is deeply rooted during this *Lifetime* comes from "family"—and usually in a direction of "authority of."

>All communication in *beta-existence* is aberrative...

...and perhaps the greatest and most intense examples of this will come from our *partners* and "elders" who like to remind us so often how "they know better" and how "you just gotta such and such" and all that.

Your present assignment then is to "*Process-out*" all of the programmed *fragmentation* associated with everything to do with the concept of CONTROL—the literal word; any time its been used as an effort toward you; any time you attempted it and were successful; unsuccessful; anything that shows up in *Processing* that would inhibit you from exercising the utmost peak of potential regarding true *Self-Honest "Control."* You can do this with any concept that you want to release programming on. You're basically working out the agreements and statements made that allowed whatever involved the concept of CONTROL to suddenly be considered "wrong"—much like what contemporary society holds in regards to the concept of RESPONSIBILITY and other key words that represent the power that is inherently yours, but of which has been disguised as "bad." The fallacy is disguised quite obscurely:

> *Don't eat from this tree—lest ye be as gods.*

Looking over our *Circles* and *Spheres*: we directly align consistently with our other *models, scales* and such only at ("0"), ("4") and ("8"). So, naturally at ("4") we have the "Sphere of the Human Condition"—or else *Humanity,* the *Homo Sapien* creatures, or *Homo Sapien Sapien*[131] depending on your preference. Our goal in *Systemology* is to elevate the state known as *Homo Sapien* to its next spiritual evolution on the *Pathway to Self-Honesty* as what some are calling "*Homo Novus.*" This is apparently a name that some in the *NexGen Systemology Society* (NSS) are using to officially classify the achieved state of *Self-Honesty*. It is the Sphere of Humanity that we are applying efforts to influence from the sphere of *Systemology-as-Organization* ("3"). The energetic harmonics are presented on this *chart* with its ripple-effects. When *Self-directed* and firm in one's resolve as *Self,* the nature of primary influence flows

---

131 **Homo Sapiens Sapiens** : the present standard-issue Human Condition; the *hominid* species and genetic-line on Earth that received modification, programming and conditioning by the *Anunnaki* race of *Alpha-Spirits*, of which early alterations contributed to various upgrades (changes) to the genetic-line, beginning approximately 450,000 years ago (*ya*) when the *Anunnaki* first appear on Earth; a species for the Human Condition on Earth that resulted from many specific *Anunnaki* "genetic" and "cultural" *interventions* at certain points of significant advancement—specifically *circa* 300,000 *ya*, 200,000 *ya*, 40,000 *ya*, and 8,000 *ya*; a species of the Human Condition set for replacement by *Homo Novus*.

outward from *Self*, rather than receiving or *being—becoming—* effects of turbulent waves. We are playing out this *Game* to essentially try to become the highest level of *Self-Honest Cause* that is possible in this existence. But, this must be *realized* equally for all those within that domain to be an effective evolution—or at least achieve a "tipping point."[132] This is why the states of *continuity-zero*, (4.0)-*beta-actualization* and (8.0) *Infinity* are the anchor points by which we generally keep *Systemology* graphic demonstrations consistent.

Beyond the inclusive Sphere of the Human Condition ("4") as a specific species-type of *living genetic beta-organism*, we essentially arrive at the zone of *All-Life* on the planet *Earth*—and that includes all *bodies* or *identities* that possess an *Awareness of Self*. This precisely means *All Animals*—but we could extend the precise definition to *All-Cellular Life*, because any form of *Life* existing on the planet with an individuated substance-form or body-structure that has a perspective—or point-of-view—from itself would qualify at ("5"). The entire *Earth-planet* is also a living body structure, and there is still some debate among early systemologists as to where it qualifies—either as a self-evident all encompassing enclosure at ("5") or as one of the many bodies composed of a structure in the entire *Cosmos* or *universe* at ("6"). At the level of ("6") we have reached a structure for *All Physical Existences*, followed by *All Spiritual Existences* at ("7"). And then of course, ("8") is *always* the qualities of "Infinity" or the "*Absolute Supreme*."

△ △ △ △ △ △

When demonstrated or effected outwardly from *Self*, these *Spheres* and *Circles* represent the equivalence of *Mastery*,

---

132 **tipping point** : a definitive "point" when a series of small changes (to a system) are significant enough to be *realized* or *cause* a larger, more significant change; the critical "point" (in a system) beyond which a significant change takes place or is observed; the "point" at which changes that cross a specific "threshold" reach a noticeably new state or development.

which is to say "true leadership" in *Self-Honesty*. This is a rare quality in our society today because everyone is bred to be a "follower"—to simply perform the task of applying our *Efforts* and personal ZU to copying-and-pasting any programming we are given. *Well, who wants to settle for that?!* Look what that mentality is doing to our planet. Do you really think the *Self* in this center spot *here* is even being actualized by the Human population at ("1")? No. *Here*: ("–1"). Not remotely *Self-Actualized*—still primarily a reactive effect of some other cause. Have you met a lot of *really Aware* people out there walking around? And its a slippery slope of "influence" with non-constructive effects spiraling down…and down…and down.

An individual down at ("–1") is now subject to their RCC pretty bad. They may be even worse off than that: with very little or no *Self-Awareness* and just running on reactive-response mechanisms. You find less *leadership* and *individuality*—more *automation* and *hive-mind*—as you move downward. The individual is becoming the invert of a leader; the invert of causes; the invert of individuality and away from *Identification* with *Self—any "Self."* Rather than exercising *Self-Direction*, such as we see in climbing upward to greater *realizations* of *Beingness*, the defining lines of individuality blur and become too extravagant and overwhelming or overpowering to be managed. So, an individual's sense of *Beingness* diminishes to the same degree as we see it expand in the other direction.

At ("–2") the lack of *individuality* or *Self-identification* leads to a lack of actualized ability or realized certainty to properly manage a domestic situation—and that personality quickly becomes antagonistic to them, including those maintained by others. You find they have a sentiment of "married people are trapped" and "people are better off alone" and all that—that is a ("–2"). Not very *Self-Actualized* individuals make a lot of "generalizations" about "people" because the individual will not see a solution to problems of domestic life, therefore the tendency becomes to exaggerate all of the "negative qualities" that can be attached to close relationships. If it cannot be *realized*, then it has to be *negated* in order to be managed—it has to be *rejected*. This, too, causes a sort of *fragmentation* that will inhibit individuals from reaching or expanding *Self* outward in the other direction to collect together and invite the qualities and resources necessary to establish these conditions.

Those who trudge through the grime of a highly *fragmented* existence and somehow manage to still reach those expressions will most likely self-destruct in view of the fact of any successes, or else sabotage their relationships. Keep in mind, this individual has already accepted and agreed to the programming that "marriage is bad" and that they are "better off alone." Do you see how that works? An individual puts a whole mess of energy into making one kind of understandable and manageable reality with a belief, agreement or postulate of some type and then some time later, thinks, "well, I'll just sweep that under the rug," or maybe doesn't even remember making this contract with *Self*, and they just go about things as if these former agreements don't exist. Well, I tell you what, you might even have some programming about "marriage" from your parents or some other authority source that isn't even your own. And if that is not effectively "*Processed-out*" then you can rest assure that this individual is going to carry these beliefs into their marriage, thinking anything can just be resolved with "a little positive thinking." *Nah-ah*.

Its quite possible that by this point you are realizing the value behind this idea of *Imagining* the *Beingness* of each subsequent "Sphere of Influence" or "Circle of Existence" as *Processing*. In performing the basic exercises, manipulating various visualizations or "thought-forms" at *Will*, you are undoubtedly going to be *resurfacing* any significant *emotional encoding* or event-based *imprints* regarding your programming toward any of these points. When you look at the constructive creative positive values on the spectrum, an individual is extending a solid reach into the environment; but at sub-levels, an individual is shrinking away and being overcome by other efforts in the environment—and of course, to equal and opposite degrees in the universe.

Practicing higher frequency thought—even in practice as an exercise—increases the flow of personal ZU, the communication of *Clear Light* throughout the *Identity* to the extent of its *realization*. That's pretty profound—yet why shouldn't that be the case? If you can get *fragmented* through some methodical application of focused attention and direction of personal energy—than why shouldn't you be able to get *defragmented* that way? All of this information seems so, "Oh, yeah. Of course." And yet, why aren't more people *realizing* a better world out there? Because we can't just place some thin sheet of "positive thinking" over a mountain of old programming and call our state "good." A lot of the more popular, better marketed, flashy, supposed "self-help" methods find very elaborate ways of doing little more than this—which all looks very nice for a minute while the check clears and before the stink comes on leaking through that paper thin napkin you tossed on top.

Now, I don't want to mislead you into thinking that this *Process* of "*Traversing the Spheres*" is a quick routine or some "fast pass" to Self-Honesty *realized* in a single session. Probably not. But the point is to practice. This is what the old mentors were trying to get their initiates to do with the *Kabbalah*—and they called it "*Pathworking*." Some modern revivals of Mesopotamian traditions do something similar with the rituals of Babylonian Star-Gate lore, and they call it "*Gatewalking*" or "*Starwalking*." It doesn't really matter what you choose to call this procedure—it is a very old and very useful technique.

As you get a sense or the "feel" about each sphere, you are tracing out a route to it as a thought-form you can interact with—so you work through them successively until you can *realize* them faster and more clearly each time: ascending higher and higher. With a little practice, you *could* travel all over the *Spheres* with very little actual effort—and in time, you should be able to essentially achieve any state—and *realize* the full range—of potential *Self-generated facets* in existence between here and *Infinity*. By achieving this effectively, you are essentially opening the internal *Gates* to the entire stores of potentiality wound up within and behind all of existence. You are—in effect—increasing all certainty regarding parameters of what is possible. This is *how* we increase the "reach" of *Self*—by increasing the extent of certainty to which there is *something* to reach for.

The divisions, ranges and domains that we describe in *Systemology* are not arbitrary. These thresholds and portions are distinguished as critical energetic check-points. For example, to exist within the range of *"beta-Awareness thought"*—what we call, between (2.1) and (4.0) on the *Standard Model* or *ZU-line*—all of that energy must be a certain type, quality or frequency to be extant at that level. Sure, it is connected to a continuous continuum of ZU energy as an *Identity*, but without this distinction of its energetic transmission, there would be no *scale*, no *differentiation*, there would be no *universes*, nothing happening for *Self* to experience. It would just be an *Alpha Spirit* or point of *Self-consciousness* existing as itself, standing as *Self*, just outside the *Gates* to an *Abyss of Infinite Nothingness*.

There is undoubtedly a very real point in existence or non-existence—it is directly relayed on the most ancient cuneiform tablet renderings described in the *"Tablets of Destiny"* book—when beta-existences were not yet formed or *imagined* into *Being*—at least before *this* version or dimension of material existence came into *Being* as (KI). These *Arcane Tablets* do not describe this activity as some random collision of forces, but as a *Self-directed* act by a single *Alpha Spirit* with the knowledge and *Will* to make it happen. There is no illusion—there is no suggestion made on these tablets, that the Anunnaki figure MARDUK[133] is some absolute *First Cause* "Creator" of

---

[133] **Marduk** : founder of Babylonia; patron Anunnaki "god" of Babylon.

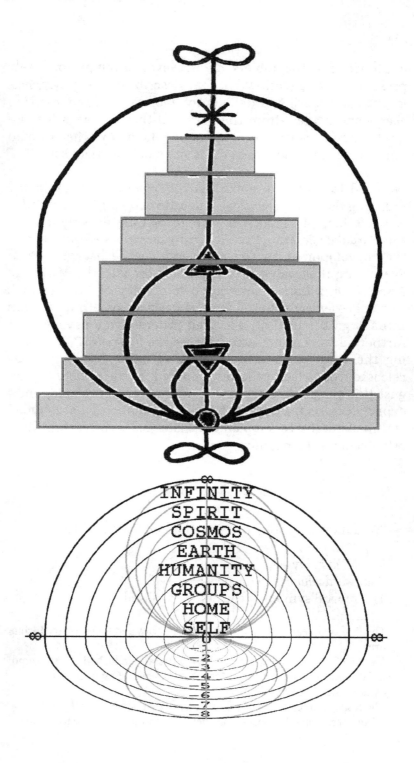

existence—no, the tablets do not even attempt to falsely present that as a truth. What they describe is a very *systematic* cosmic ordering, intelligently fabricated from existing *energy* and *matter*—from the potential that *is*. There is absolutely no reason why *realizations* held by the Human Condition should not reach these same heights *right now*.

We used to talk a lot about *Reality Engineering** back in the founding days of *Systemology* a decade ago—until I realized it was too steep of a launch point to bring folks directly to from the *Mardukite Core*. So, I spent eight years *systematizing* its further development on the back end—in the background of everything that was still visibly going on with the *Mardukite Research Organization*—and reached a point of high enough certainty—*for my self*—to bring us back to something we can actually grab a hold of, and then progressively develop into further advancements. And here we are in *Grade III*—continuing the work outlined in *"Tablets of Destiny"* just recently released from a lecture series prepared for the Society∞ in August this year. The work for *"Crystal Clear"* completed, and I expect lecture transcripts from this month to be included within the book to supplement materials drawn from these other reports‡ I am issuing to you.

△ △ △ △ △ △ △

> "And more, my son, for more than once when I sat
>   alone,
> *Revolving in myself,*
> *That word which is the symbol of myself,*
> The mortal symbol of the Self was loosed,

---

* The title of an early *Systemology* lecture series and textbook by Joshua Free.

∞ *NexGen Systemology Society (NSS)*—officially launched underground in 2011 as a division of the "Mardukite Research Organization" (Mardukite Chamberlains).

‡ The tests, chapter-lessons and *processing* that appear in *"Crystal Clear"* were first dispensed confidentially to the *NSS* in October 2019.

> And passed into the Nameless, as a cloud melts into heaven.
> I touched the limbs—the limbs were strange, not mine
> —and yet no shadow of doubt,
> But utter clearness and through loss of Self,
> The gain of such large life as matched with ours
> Were suns to spark—unshadowable in words—
> Themselves but shadows of a shadow-world."
> —*Alfred Lord Tennyson.*

It is a strange and esoteric sentiment to say that an individual has "found themselves" or that we are on a Pathway of Self-Discovery. I mean, who is doing the looking and discovering if it isn't the *Self*? It would seem to me, that the most aberrative effort a society could ever do to maintain control by confusion and fragmentation, is to tell its population of Humans that they need to spend the rest of their lives "finding themselves." Well, you can go and occupy your time chasing your tail if you like or looking for the glasses on top of your head—no one will begrudge you your *Self-governed* "right" to go and do that if it makes you happy. But at least *know* that you *are* doing it. We all like to play *Games*—it is what occupies the *Alpha Spirit* at the core, underneath all of these layers we have added to *Self* that we attribute to our *Identity*. But that's just it—because what happens when you don't *know* you are playing a *Game* is you are a pawn—an *effect*—you are allowing yourself to *become*, to put your *Beingness* into the result of control by someone else. Period.

[*Audience murmurs.*]

Look: that's why we are calling this segment of *Grade III*, "*Crystal Clear*," because we are cleaning up the lenses now, clearing away the *fragmentation* and giving it a good looking through at this thing we call *Life*, *Reality* and *Existence*—this thing that we call "ZU" on the *Mardukite* end of this. It's not always pretty at first, but its the truth of things, and its what we are here trying to clear up at this point of development. It would be hard to *Self-Actualize* much further unless we can isolate this most occult and esoteric of all things between here and *Infinity*—this thing called "Self." And for those of you that prefer themes with conical hats and wands: the individual *Self-Honest* and realizing the full potential of the *Will-Intention* at (5.0) is your "Tarot Card Number One"—*The Magician*. If that is your aspiration, than this *Pathway* we are illustrating in *Systemology* will work just as effectively to your ends as well. *Systemology* is *universal*.

WILL is what encompasses the first order of understanding in the spiritual existence. You see, everything you associate with the ZU-line moves up and compresses as a "universe"—that's why there is so much more wave action going on in that band up there. WILL is fundamentally the spiritual equivalent of *emotional energy*, but up in the frequency of spiritual existence. It is the *Efforts* applied directly by the *Alpha Spirit* to establish a manifestation of *Beingness* independent of the "I" or *Self*. You are, after calculating—or using trial-and-error about—the *Logics* and *Games* of universe mechanics, putting that pure expression out as WILL. It is probably around here (5.0) that the ideals of *Systemology* and what some call "magic" *do* actually meet. The difference is the actual potential realized as that expression while operating in *beta-existence*. There are not a lot of "magical paths" out there that lead directly to a point of *Self-Honesty*. Even though they may produce some kind of results following basic principles of Cosmic Law in applying personal WILL, it is not at all the same.

Therefore, as a static point for examination:

—the point of WILL at (5.0) is the potential of actualization by a first order of understanding in *beta-existence*.

And that is only potential, that does not mean its the same for everyone. On this same logic:

—the point where *Games*, *Logics* and other *Universes* are realized into being (6.0) is a potential for the second order of understanding. This is regarding the *wisdom* and "true knowledge" actualized from from the second point of realization in *beta*; and finally,

—the point of *Alpha Spirit* at (7.0), which we consider *Total Self-Actualization* as an *Alpha* state, is our aim from the third and highest order of realization and understanding achievable from "communicable knowledge" and use of language. This is achieved only on *Self-direction* of an individuated *Self* with all of its channels of communication *defragmented* up unto this point. They should be able to handle the rest from there.

The reason differences between *realization* and *actualization* are so firmly impressed is because we are dealing with *Alpha* and *Beta* existences linked on a ZU-line of *Self-Awareness* and that this word that we throw around—"potential"—has a way of getting people "loopy" in considering the angles from opposing viewpoints. Which is then also why we stress being "in phase" from the perspective of the "I" at all times, because it is too easy to fall out of that "seat" and *remain* out of it.

The "I-AM" exercise I suggested in this lecture can be best realized not by simply leaving the page of "I-AMs" blank—because that would not be *Self-Honest*. You can do it, but you aren't impressing anyone if its not true for you. The real trick is not ignoring the *facets* attached to *Self* as a "personality," but by the dissolution of it. *Dissolving it.* This is what you are doing when you rise up in actualized *Awareness* or traverse the *Spheres* with your Mind's Eye. So, the trick is then to consider another page to add to this—this wasn't how it was given to me back in my "psych class" days, but it is something that I decided to experiment with later. You take your list of I-AMs—you'll want to do this in pencil; everything in pencil when pertaining to scribing of *Systemology* paperwork—and so you have all these things here on your palette, and then you take this other piece of paper and you put "NOT-I." These fulfill the entire "esoteric monad"—as they call it—of the "I-NOT-I."

Alright, so you take your list of "I-AMs" and you just start *Processing* them out, one right after another. I mean, I can fill a book of all the possible ways a person might be *fragmented*—and maybe at some higher up Grade of our materials that might be useful as a reference—but the point of the "*Crystal Clear*" volume is to take a *universal systemology*—an objective holistic spiritual map that has been found workable on its own—and provide an individual an effective route for the *Pathway* that is applicable to the *individual Seeker*, on which they may then tread with certainty. This means that for this exercise, you take all that which you have accumulated and *process* it "out"—thereby you are able to run your "eraser" over the word, term or association a single time. Every time you "run that out" a little more in a session, you erase it a little more. When you have no reactive-responses or programming inherently attached, erase it completely and add it to the "NOT-I" list. We may even later go ahead and add such worksheets to the official *Adventure Journals*[*] just for a *Seeker's* convenience.

The witching hour of Halloween is closing in, so we will close this lecture series with this important consideration: With this *Processing*, you are shedding perceived layers of "My-Self" that have been accumulated and have been eclipsing the "I" to a point where it is often no longer even recognizable. It is for *that* reason alone that we call this *Self-Discovery* at all. Because in the end, when you stand there stripped bare at the point of *Self* again, it will be like meeting YOU for the first time. *Fly straight*—and have a happy and safe Halloween. Thank you.

---

[*] "*Systemology–Seeker's Adventure Journal*" Coming Soon (2020).

I AM

...

NOT-I

...

UNIT FIVE

SELF DIRECTION

"I WILL, THEREFORE—I AM"

Descartes* once wrote: "*I think, therefore I am*"—and suddenly a modern pursuit of *epistemology*,[134] which is really the most ancient philosophical pursuit, sprung back into being. It remained alone in the field of philosophy because it could not be adequately systematized using physical sciences. Therefore, society has never been able to definitively establish *proof* of Human understanding of anything—only a shadow of truth which is contained in a continuity of vocabulary and semantics applied to causal reason and logic and cosmic ordering (sequence) that is observed within an acceptance of that *beta* paradigm only.

Modern "Psychology" made an attempt to break away from the *philosophical schools* to become an empirical/physical science, but it failed in its original mission, purpose and function to *systematize the Mind-System* and *consciousness*. And as an entity, in the "Third Sphere of Existence" (*organized social systems/"groups and societies"*) it failed in its *Efforts* to actualize this for all humanity. *Fragmentation* set in; a downward spiral of *Awareness* began. Now we are left with an uncertain physical science of the physical brain. The fields of mysticism, magic, spirituality and traditional religion all tried to establish a firm hold on the "Third Sphere of Existence and Influences" yet used techniques and methods that could only be realized in the "First Sphere" unless otherwise directed by enforcement and threat of punishment. This is where most of the fragmented *Human Effort* of the last several millennium has left the state of things.

Former lesson-chapters subsequently worked through an understanding of basic methods a *Seeker* might take—by themselves unassisted—to *realize* cumulatively higher *actual-*

---

*　*Rene Descartes (1596–1650)*, french philosopher and systemologist.
134 **epistemology** : a school of philosophy focused on the truth of knowledge *and* knowledge of truth; theories regarding validity and truth inherent in any structure of knowledge and reason.

izations of *Self*. There are numerous additional applications that demonstrate *Processing* at each of these former domains—but here a *Seeker* is given a crash-course on fundamentals. These are sufficient enough to develop subsequent applications using existing materials as given in this volume with each *Seeker's* "cycle" or pass through the *"Crystal Clear"* Self-Defragmentation Program—of which it is assumed you may require several cycles of working through this "workbook" (quite appropriately named). At each level of holistic application, the way forward on the *Pathway* all inclusively takes on many styles that seem reminiscent to other associated methods: appearing at times like an exercise regimen; at others, a day in the life at a spiritual retreat; and then, when we start working with WILL at (5.0) it starts to seem a bit more like "magic" because we are working on a level just across the "veil," exterior to the (*Beta*)-Physical Universe-(KI) that we identify any frequencies in.

The application of personal WILL on the *ZU-line* at (5.0) is beyond, outside or *exterior* to the Physical Universe (KI) and all measuring devices used in *beta* to calculate expressions of ZU. It is beyond even the "thought waves" that are measured on an EEG—which is only an approximation for graphic calculations of "beta-consciousness activity" between (2.1) and (4.0). There are "biofeedback" devices that can evaluate "state-change" as it applies to emotional reactive-responses under (2.0) when the RCC is directly engaged—but again, there is *no* known device or means within our technology that pertains directly to WILL; only the *effects* of WILL as it travels down the frequencies of a ZU-line in *beta-existence*. This is why, if Cause is not *determined* to be *Self-determined*, we can never be absolutely sure *what* the nature of any Cause is—that is to say, regarding the world we are seeing and interacting with as a Physical Universe (KI).

> WILL (5.0) is *Self-Awareness* of directing *Alpha Thought* (6.0) into a creative expression (below 5.0).
>
> The *Self* as *Alpha Spirit* (7.0) is the "Cause" of WILL (5.0).

> WILL (5.0) is the *Self-Awareness* of Potential
> Infinite-Manifestation of the ALL (7.0)
> *realized* as *beta-expression* under LAW (4.0).

The *Self* is only the Prime Cause:

a) when in its inherent natural-spiritual state, fully actualized; or

b) when the ZU-line connecting *Awareness* to a *beta-existence* is completely *defragmented* in *Self-Honesty*.

These are no slight conditions—but they *are* attainable.

The first (a): *actualized* while *exterior* to the "*body*"—and

The second (b): *actualized* while occupying a *beta-existence*.

The second is achieved while placing focus on the first.

Realizing the first will *Actualize* the second.

For the *Alpha Spirit* to actualize its own expression as a condensation down the ZU-line, it must be *Self-directed*. Any form of *beta-fragmentation* will greatly inhibit this free and total expression in the Physical Universe (KI). For one: it will not be fully actualized with full *Effort* down the chain in *beta-existence*, if it is not clearly expressed at its inception (as *Will-Intention*). There are other forms of *fragmentation* even at the *Alpha* levels of existence, but for us to "Return to Source" via our ascent back up the *Pathway to Self-Honesty*, we are most concerned with the *fragmentation* that has occurred either in the current lifetime that the *Seeker* is occupying—or if possible, even including the "genetic memory" that is stored in the *genetic organism* even prior to be commandeered by the *Alpha Spirit*. Once the way has been cleared through all *beta* levels, only then should a *Seeker* start to apply attention directly to any *spiritual fragmentation*. It can otherwise become quite enamoring and pose a distraction away from actually achieving the necessary *beta* results ahead of it.

A strong, clearly managed, *Self-determined* WILL (5.0) is the fundamental key that underlies all spiritual, mystical and magical *Mastery*. It should not be a surprise that *Self-Mastery* and *Mastery* of all of the other "Spheres of Influence and Circles of Existence" are to be attained via the same route. Thousands of years of esoteric and mystical renderings already have suggested this to us—but former use has not necessarily led to solidity as a true *realization*. This is one of the reasons why restricting pursuit of *Self-Actualization* to any former methods emphasizing one mythological or conceptual paradigm in exclusion all others has not provided optimum effects. This is to be the case especially now in the 21st Century when we have access to the most ancient and holistic demonstrations of *Mardukite Babylon* and *Mardukite Systemology* at our disposal. There is no going back now that these new heights for the Human Condition have been *realized* for the future.

The highest point of *realization* that we are considering in the "logic" of *Systemology* is the point of *Self-Actualization* as the *Alpha Spirit* able to *realize* itself along the entire ZU-line of *Awareness*, while still remaining in-and-as *Self* in full *Awareness*. This degree of *Actualization* is achieved at (7.0) in the Spiritual Universe (AN). It may, however, be *realized* into *Being* for consideration by *Self* in *Awareness* at any point of evolution. This is one of the ways in which *beta-defragmentation* processing employs the highest frequencies of *beta-Thought*. What does this entail? It requires tracing origins of whatever is being "processed" through the WILL—the entire *chain* realized back-to-back as an infinite continuum. This is a natural ability of the *Alpha Spirit*—to monitor activity of ZU along the entire personal *chain of causality* into *beta-existence* (or even an intention of creating a *beta-existence* or Physical Universe) without simultaneously fixing an *Awareness* to it and becoming a permanent *Effect*.

SELF-DIRECTING WILL AND INTENTION

The WILL (5.0) applies *intention*, but the directive *Alpha Thought* (6.0) and subsequent *beta-Thought* are doing all of the analytical, logical, intellectual "processing." *Intention* and *Emotion* are two ends of a spectrum in the continuum representing personal "force" and "effort" exerted on—or imposed on—the Physical Universe (KI). *Defragmentation processing* clears energetic communication channels between *Alpha Will-Intention* and *Beta-Effort*. This communication takes place in the band of thought associated with the "Mind-System." Therefore, as an intermediary between Spiritual (AN) and Physical (KI) existences, the *Master Control Center* (MCC) is the critical component that evaluates, analyzes and determines the type-quality and intensity required to cause a desired *change* or *effect*. This is all determined in the "Mind-System" based on information it is "processing." And a *Seeker* realizes by this point that quality and accuracy of the information is entirely related to conditions of personal *Awareness* maintained and degrees of "true knowledge" managed. Every single *facet* or *element* that we could possibly *think, feel* or *do* is a result of this information. We cannot understand any *facet* or *element* that is not first *realized*.

The basic functions of the "Mind-System" all relate to one or another steps regarding:

- Critically Analyzing a Problem or Condition
- Applying and Communicating Energy/Data
- Evaluating Results of Applied Personal Efforts

Although it is easy for the *maladjusted* Human Condition to present problems as some wide arrayed mess, the solution to them, and the application of efforts, results from a series of small "YES–NO" answers delivered as a systematic arrangement of much smaller *facets* and *aspects*. In essence, much like a computer does a vast array of complicated and elaborate functions, the actual processing behind it is a series of "ON–

OFF" sequences that we tend to refer to as the "*ones* and *zeroes.*" There is no gradient in this part. Even matters of "intensity" are simply evaluations of how much Effort to apply. If you consider the attempt to "crack a safe," the application of each combination is nearly equivalent to the *effort* of each other combination until the procedure is repeated long enough and with enough gauged variety to produce the desired result. You might notice how the same idea could be easily demonstrated in "testing the physical integrity" of an object. For example, if you were to hold a wooden pencil at each end—with your thumbs against it for leverage—and begin to slowly apply pressure to it, you would go through a series of instantaneous cycles of thought: "a little more?" -*yes.* "a little more?" -*yes.* "a little more?" -*yes.* — SNAP!

[*subjective-processing example*]

—RECALL a moment when you *decided* to *push* against something.

—RECALL the *effort* you applied to actually *push* the object.

—RECALL an instance when you were *pushed* by someone else.

—RECALL the *effort* you applied to actually withstand the push.

[*objective-processing example*]

—LOOK around and select an object you could move.

—IDENTIFY the object and then *you* touch it.

—DECIDE to apply the *Will-Intention* to move the object.

—APPLY the *Self-Command* that: "You move the object."

—DECIDE to apply the *Effort* to move the object and then *you* move the object.

For all the emphasis and sentiment the Human Condition is programmed to attach to the *past*, there is only one valid, useful, worthwhile use of mental faculties of a defragmented *"memory"*—and that is "estimation of efforts." We use previous knowledge of *efforts* that *Self* has applied in order to determine future *efforts*. This is not generally a source of *fragmentation* in and of itself. Most of the *fragmentation* that affects us most destructively comes from *efforts* others apply to us and the manner in which these greatly affect programming that contributes to later *fragmenting* an estimation of our own *efforts*. Solving problems is a matter of *certainty* combined with *correct* information. Whenever energy and/or communication is *fragmented, withheld* or *enforced*, the amount of *certainty* and *information* will be diminished along with the corresponding level of *Awareness*.

Examine just a few examples affecting you now in your actual life by listing the first problems that enter your thoughts or that have been bothering you, which have not yet been dealt with (or which have not yet been solved):

    A. _____
    B. _____
    C. _____
    D. _____

It is likely that these lingering problems remain unresolved as a result of *fragmented, withheld* or *missing information*. If you find that this is the case, write down the information that would resolve the equation/problem:

    A. _____
    B. _____
    C. _____
    D. _____

With these considerations in mind, decide whether or not new information is still necessary to solve the problem. If it is, write down how you might go about determining this information. If new information is not necessary, decide whether or not the problem can be resolved with *Analytical Processing*, and write down those results. If you have already determined the course of actions required to solve the problem as an *effect*, than write down the *Will-Intention* required.

    A. _____

    B. _____

    C. _____

    D. _____

Many "problems" that the *Seeker* will face in this lifetime are a part of basic *Game-conditions* inherent when *Self* interacts with the Physical Universe (KI). The type of problem-solving that is treated in the Spiritual Universe (AN) relate specifically to conditions of the *Alpha Spirit*—and these relate to the highest order of *Logic* evaluated by *Self* in its direction of *Will-Intention*. Once an intention has been set forth, it is up to MCC systems to enact the *effect* in *beta-existence* as an intermediary "cause." Communication of energy is relayed along down the ZU-line in this fashion. There is also another type of "problem" that is erroneous and generated as a result of *fragmentation* and disrupted *programming*, such as "confusion" or the "counter-effort" of others applied against us. These may appear to be very real problems when *Awareness* is fixed to them by emotion, but most often, these can be resolved with *defragmentation*. This provides the most permanent and recognizable results, assuming some other method might be applied to get an individual momentarily unstuck.

If a Primary Decision (*Alpha*) can be actualized that a "problem is non-problem," than the result will be the same. If, however, an agreement has already been made that it *is* a "problem," than the *Seeker* will need to "*Process-out*" or dissolve the original agreement/belief through to its entirety. One cannot label and reinforce something as a "problem" *and then* decide that "oh, well, I guess I won't deal with it." That will not resolve the issue and the issue will resurface later as if *imprinted* with *emotional encoding*.

Many individuals superfluously accumulate an entire museum library full of *beta-problems* that are actually non-problems, but for whatever *fragmented* reasons, the individual is following programming that seeks to *use* this accumulation as a quality of its *beta-personality*. This, too, should be "*Processed-out.*" The alternative is holding onto old erroneous "problems" as *solids* and *facets* that later affect the solving of problems in the *future*.

There are probably more than a few problems with environmental situations, other individuals and physical masses that you faced in the *past*—problems that you very seriously considered "problems" and which remained unsolved—or remain unsolved—to this day. Following in a similar mode as previously instructed, list the first such unresolved problems that come into mind:

    A. _____
    B. _____
    C. _____
    D. _____

In the same mode of treatment used toward present concerns, write down any *fragmented, withheld* or *missing information* that you could have used to resolve this past equation/problem if you had known it at the time.

    A. _____
    B. _____
    C. _____
    D. _____

With these considerations in mind, decide whether or not this new information has since surfaced to resolve past concern. If it has, write down how you determined this information. If the necessary information was never disclosed, decide whether or not past concerns can be resolved with *Analytical Processing*, and write down those results.

A. _____
B. _____
C. _____
D. _____

If necessary, *Process-Out* lingering fragmentation. You may draw a clean line through the notes indicating whenever a *present* concern (a problem repeatedly a current focus of attention) or a *past* concern still actualized in *present time* (a lingering *imprint* or *emotional encoding* that has fixed *Awareness* to *past* failure of efforts) is deemed a "non-problem" or "*Processed-Out.*" The experience of resurfaced *facets* and *Analytical Recall* of moments when you have successfully solved problems and effected results in accordance with *Will-Intention* should also provide greater certainty of *Self-direction*.

RETURNING THE REIGN OF SELF-DIRECTION

Whenever we apply personal *efforts* toward our *beta*-management—of the body, of the immediate environment (Sphere of Influence), and the Physical Universe (KI)—we are *Self-directing* "control" of personal attentions and energy as WILL (*and intention*) toward accomplishing a goal, mission or task. These proposed "goals" become our "purpose" during our lifetime—for there is *no other* beyond the "Prime Directive" to *exist*. However much these other goals are self-imposed, they are not always *our own*. We take on goals in agreement with *imprinting* and *programming*—and very often the source of such goals, attitudes and missions of our lifetime—that drive us on—have been implanted, inherited or assumed from failed goals of loved ones, departed ones and supposed authorities. We've all seen classic examples of parents enforcing a certain career goal or lifestyle choice onto their children—but this is only the most obvious and overt type of imposition easily and readily recognizable—because we are at least consciously *Aware* of such dynamics on the surface.

What about the goals and driving attitudes behind your actions, which are based on *programming* that you are not as *Aware* of—and which you cannot as tangibly stand up to in present circumstances? Consider the goals and attitudes assumed by a soldier reared to fight, kill and die for a cause he doesn't understand. That lack of *certainty* must be compensated with sheer *conditioning*—forget even traditional programming and education, because a military prides itself on an army of clone-robots that will only react on command and can be expected to process and follow command without exception or thought.

> The goal of *Self* for actualization and the goal of an enforcing commander are the same: instantaneous effect by reduction (or removal) of all communication lag. Once WILL has been directed, the ensuing motions of the system below (5.0) are expected to "snap to it" immediately and without "thinking"—assuming they are appropriately *defragmented*.

Awareness goes from *Be* to *Do* on practiced command and without hesitation. This type of reconditioning has been found to be most efficient and provides satisfactory optimization for *beta-Awareness*. We are not concerned with any "worldly" semantics used to describe these methods —because an individual working alone could just as soon employ these same type of *drills* for themselves and attain quite effective personal *Self-discipline*. Therefore, our concern is only —as always—*who* is in control? Is it *Self*? And is *Self* operating *Self-Honestly*, independent of all imposed *conditioning* and *programming* from any "outside source" as "cause"?

In the previous section the *Seeker* is asked to identify *present* and *past* problems—to the extent that they are perceived to be problems. But—

> ...an *unresolved problem* is simply an *unanswered equation*.

It is simply waiting for you to *make a decision*. It only remains in the realm of "problem" so long as there is *uncertainty* about it, which is always traceable back to a source of *fragmentation*.

Obstacles, barriers and challenges are *Game* conditions of this existence. These things are the way they are because *someone* has put them in motion *to be* as they are—even if from an outside source that is completely independent of the *effect*. And *that* is where we assign "power" to. The person that is telling you how you should be, what you should do, and how you should do it—and yet they, themselves, are impervious and independent of any effect of these things—that person is attempting to be your *Alpha Spirit*; to essentially assume control of and commandeer your *beta-existence*. And these same individuals, running on their own *fragmented programming*, are generally the perfect example of everything *not* to be and do in this lifetime, demonstrating that they are incapable of handling and managing their own physical and mental and spiritual responsibilities adequately, so they reinforce the security and certainty of *their* Reality by enforcing it on you. As soon as you fall into agreements with their position, you suddenly have a new *god* directing your existence that is not *Self*.

An individual's certainty and contentment with their existence is closely tied to the problems that are solved. The true individuated *Self*—the *Alpha Spirit* that is directing ZU from the Spiritual Universe (AN)—is actually quite fond of, and well equipped to, create *Games*, play *Games* and solve problems in the Physical Universe (KI). It is, in this sense, a *god*. It is able to effectively do exactly what others are attempting when they try to control us: it is *Self-directing* the *Will-Intention* and *Effort* from a position that is *exterior* to the *effect*. After information is sent back along the *ZU-line* (hopefully clearly communicated) concerning results or *effects*, additional educated and calculated *Will-Intention* and *Effort* may then be applied. This is the *only* time we "learn" anything from experience—and even then, what we learn is only specific to the exact conditions for which we applied the *Efforts* and discovered *Effects*. These conditions—or the information regarding them—may very well be *fragmented*, which is important to know about because it will, in turn, affect how an individual approaches *all similar future problems*.

We have discovered that an individual's *certainty* and *Self-direction* toward accomplishment of future goals, overcoming future challenges or barriers, and solving future problems, are all linked very closely to the type of *beliefs* and *imprints* maintained about all *past* and *present* successes and failures—particularly the most recently experienced ones.

All failed and unrealized goals become a source of *fragmentation*. This necessarily includes the goals and purposes that we decide to take on from others. We also have certain *expectations* concerning results of our *Efforts*—and when these are misaligned or met with *counter-efforts* from others, the outcomes change what we are seeing (or how we are sensing incoming information) and interacting with the environment thereafter.

The nature of the relationship between the *Self* and the environment (or another person) changes or is adjusted to accept the new information. This is how we develop the experience in *beta-existence* that contributes to our "assumed personality" of likes, dislikes, inclinations and tendencies, &tc. Very few of these, if *processed*, will be found to be the *actual* aims and goals of the "I" or *Alpha Spirit*. And there are more than a few of these that an individual carries with them that are just purely aberrative.

For example: a young man carrying around failed realizations of goals put forth and emotionally reinforced in youth, to "grow up famous; make lots of money to support his parenting generation; buy his mother a home; become a dentist like dad; take over the family business; go pro in baseball like uncle so-and-so asked on his death bed..." and so on and so on. Of course, there is nothing inherently wrong with any of these sentiments—but when we carry them around with us as *failures* or imposed goals, they become sources of *fragmentation*. When we are worrying and placing anxiety on present concerns due to past failures and erroneous *imprinting*, we are not "in phase" with the present—and this is one way we calculate, evaluate and assign the "numeric values" that we do to the various states of *beta-Awareness*.

For the present author to scribe enough *Processing Command Lines* to make this portion of the book effective to all *Seekers* objectively would require an entire volume, or several, for that purpose alone—of which the majority of it would be considered non-applicable for a general case in every other instance. The purpose of supplying education and examples is specifically to direct *Seekers* to consider instances that *are specific* to their own experiences. Of course, to leave a *Seeker* in "free-association thought" for this task would not be very *systemological*—so, as in the prior section, the *Seeker* will be directed to either write in the spaces provided in this book (or a companion notebook/journal) for evaluation and future reference. Keep in mind that a *Seeker* is expected to work through the whole cycle of materials presented as *Crystal Clear* additional times. The purpose of this should become most effectively observed, self-evident and enlightening on the *third cycle* through, when a *Seeker* will find that they are handling and managing *personal fragmentation* with far greater ease than before, until it can be simply *dissolved at will*.

Make a list including the most basic and elaborate "goals and ambitions" that come to mind, which you have *realized* for your life and that you are still working to *actualize* in the present (and future).

1. _____
2. _____
3. _____
4. _____
5. _____
6. _____
7. _____
8. _____

Now consider all of the different people that you have encountered in your lifetime, which you consider to have been "close" to you physically and share "affinity" with you emotionally—such as in the second "Sphere of Influence" (*Home, Domestic Life, Family*). Take a moment to consider all of those of whom you have shared this "closeness" with—and which specifically have since departed from you, either as a result of death or permanent disconnection. Place these names on each numbered line.

1. _____

2. _____

3. _____

4. _____

5. _____

6. _____

7. _____

8. _____

Take a moment to consider any unfinished aims, goals, fears and dreams—*realized* but not bought to *full actualization*—for each one of these *lifeforms* and list them with the names. These are the *failures to actualize* that the *Self* has perceived for each of these close individuals.

> If you find that any of these *failures to actualize* are similar to your own present or future goals and purposes, circle the entire name section and make a note of the relationship on your own personal goals list.

If you find that RECALL of any of these individuals or their relationship to you causes you to have a deep emotional reaction or distress, then *process-out* the entire relationship with this individual to the fullest analytical extent. Do this until you are comfortable in recalling memory of these individuals without stressful reactive-responses.

[*Identity-Phase processing example*]

—RECALL the first memory of when you met ___ .

—RECALL the first instance you made the decision to *like* ___ .

—RECALL any moment you made the decision to *like* ___ .

—RECALL the first instance you made the decision to *be like* ___ .

—RECALL any moment you made the decision to *be like* ___ .

—RECALL the first instance you *realized* that you *were* like ___ .

—RECALL any moment you *realized* that you *were* like ___ .

—RECALL the first instance you made the decision to *be sympathetic* toward ___ .

—RECALL any moment you made the decision to *be sympathetic* toward ___ .

—RECALL the first instance you *felt anger* toward ___ .

—RECALL any moment you *felt anger* toward ___ .

—RECALL the first instance you *regretted* something you did to ___ .

—RECALL any moment you *regretted* something you did to ___ .

—RECALL the first instance when you tried to *help* or *assist* ___ .

—RECALL any moment when you tried to *help* or *assist* ___ .

—RECALL the first instance when you successfully *helped* or *assisted* ___ .

—RECALL any moment when you successfully *helped* or *assisted* ___ .

—RECALL an enjoyable event or memory you shared with ___ .

—REPEAT the above processing cycle again. (*Repeat it several times.*)

—RECALL the instance you started this processing session.

—RECALL the instance you completed the last cycle of processing.

—RECALL the most recent time you enjoyed what you were doing.

In traditional esoteric mysticism, there are various "oracles" consulted to "divine" information that is otherwise hidden from conscious *Awareness*. One of the key areas of query used to determine a general state or any overall outcome is very literally the "hopes, dreams and fears" of an individual. These are equivalent to, if not greater than, the importance of the *Will* and other *beta-Awareness* taking place to produce an *Effect*. The reason these seem to be more critical than many other types of *emotional energy* and *thought-forms* is that the "hopes, dreams and fears" concerning personal *certainty* and *Self-management* will start to affect beta-circulation of the energy coming into manifestation from (5.0) WILL.

The type of *imprinting* and *programming* that targets the "hopes, dreams and fears" of the Human Condition is, perhaps, the most *fragmenting* of all experiences in an individual's lifetime other than the matter of a violent death. You could change a person's *beliefs* and affect their personality and tomorrow they might change them again and take on a different personality—but when you start invalidating the *hopes* and *dreams* of humanity; well, just start prepping mass-graves. Of course, this is all reversible—as are its effects—so long as the Human Condition is *willing* to take the *responsibility* and attain *true power* of the kind that has never been actualized on Earth for thousands and thousands of years.

SELF-DIRECTING BEYOND GUILT AND BLAME

When a *Seeker* is directing personal *Will-Intention*, they are *Self-directing* the ZU present in *beta-existence*. All *Self-directed* activity conducted by the Human Condition begins first in the realm of *Spirit*, the Alpha existence of the Spiritual Universe (AN), where resides the true embodiment of the "I" as *Self*. The nature of ZU energy—or rather its type and frequency—is what others are sensitive to when they "pick up on" any communication or information that is not directly connected to spoken words. Even when we do speak words, it is often not the words themselves that carries the greatest intensity of our communication. Some refer to an emotional quality of the sound connected to words as "pitch" or "tone," but whatever we choose to call it, it is felt directly at emotional levels far more than intellectual ones—and it is often felt to such an intensity and solidity that we almost tend to associate the force directly with the level of the Physical Universe (KI). Deep emotion is *thick* and *syrupy*—often forcing more and more *Awareness* directly into the *physical/genetic body*—and far from the type of elated/elevated vibrations experienced in higher-frequency thought.

Emotional encoding, *conditioning* and *imprinting* are all heavily charged emotional forms of personal *fragmentation*. Since our personal *emotional* vibrations are also tied very strongly to our "output" and evaluation of *Effort*, it stands to reason that at the highest level of reasoning: *Emotion*, *Effort* and *Intention* are all very intimately interconnected systems. These systems are ruled by various degrees of thought and consideration. The purpose of the *Alpha Thought* system is to "create" or "control a creation"—or communicate a directed command to do the same. It is the responsibility of other *control centers* to receive commands and issue the actions. All of this runs quite smoothly, is clearly thought and imagined— and properly executed—so long as there are no significant sources of *fragmentation*.

Beyond matters previously discussed in this present volume, the next most important aspect to *Process* and manage regarding *Self-direction* is actual "assignment of responsibility" that we have concerning ourselves, our environment and aspects of the Physical Universe (KI). Each of us has the ability to create and dissolve our creations; to decide and to change our decisions; to act and learn and then change how we act the next time—each of these qualities only belongs to the *Alpha Thought* of the actual "I" (*Self*); of the *Alpha Spirit* alone and no artificial *beta-personality*. We have incredible power to control, and capability to dissolve, manage and change anything that is actually our creation and responsibility. When the assignment of responsibility, ownership and control is misaligned and misappropriated to a source other than what it is—only then does it become exceptionally "solid" and unmanageable. This is one of the great tricks of *fragmentation* behind the structure of the material systems in modern society: Get all individuals to treat problems as someone else's responsibility, and they will never go away.

When we consider the solidification that takes place on the ZU-line—particularly in terms of *imprinting* and *thought-forms* —the only *fragmentation* a *Seeker* can easily resolve, especially when performing *Self-Processing* alone, is what is most readily accessible and "owned." It is far too easy to pass off blame, responsibility and ownership for all of creation, all of our *imprints* and all of the problems in the world—but then we lose all power to effect a *Self-directed* change in *beta-existence*. Of course, it would be just as *fragmenting* to go out and take on *all* the responsibility and blame for the world's problems. Somewhere in between there is a healthy balance or middle ground that reflects what is within our ability—and direct responsibility—to actually take *Self-honest* "credit" for "creating," and then manage accordingly. And the one thing we absolutely *do* have a hand in creating, agreeing to and taking on like layers of clothing, is this thing we call *"personality"*—or, if you wish to be more specific: the *beta-personality*—which is not the "True Personality" carried by the *Alpha Spirit*.

At the intellectual level of "false appropriation of cause," we find low-level thoughts of "blame." And when we take that frequency down into emotional circuits, we discover "guilt." There is nothing wrong with taking responsibility for *Self*—taking measures to *defragment* programming and response-actions of *Self*, measuring and evaluating consequences, and so forth—we encourage that because it yields effective results on the *Pathway to Self-Honesty* and beyond to the total actualization of *Self*. But, what about the emotional solids that we carry around with us that are connected to misplaced responsibility and misappropriated causal observation? These do not simply "go away" unless they are *Willed* away appropriately. If not, misappropriating the cause and sources, we attribute Power of "direction" *someplace else*.

Without "true knowledge" of *cause* and *creation*, we cannot expect to maintain *control* of it. It remains someone else's *cause* that we agree to take on as a load to carry, wear or assimilate—but it becomes less effectively our own *creation*. By misplacing identification of responsibility of other beings onto ourselves, we lose the clarity necessary to fully actualize or clearly *Will* anything *Self-Honestly* and *Self-directed*—because we will be using false information to make these judgments. It works the other way as well: the *Imprinting* that we have *created* ourselves is entirely within our control to alter or destroy at *Will*—they are our *creations* during this lifetime that we *have* participated in. Well, as soon as we pass *these* off onto others and misappropriate *other individuals* as our cause, we are reducing our own *Awareness* on the responsibility of managing *Imprints* and *Programming* that *is* entirely within our realm of *control*, should we wish to resume the reign.

Using *Systemology Processing* to "*Process-out*" and sufficiently *analyze* and disperse emotional ZU-energy wound up in "guilt" and "blame" is an important step for actualizing WILL (5.0) in *Self-Honesty*. So long as a *Seeker* has actualized enough *beta-Awareness* (high enough personal frequency) to confront the nature of guilt and blame, an application of basic AR-SP-2 should suffice in resurfacing significant emotional solids for reduction by *analysis*. In this case, the *Seeker* elevates to a high-frequency of *Awareness* focusing on (recalling) certainties, positive/successful experiences and then afterward bring out the times of stress regarding responsibility, blame, guilt, shame and regret. In most cases, the emotional mechanics will be found to be *fragmented*—an appreciation of which can only be attained at the *analytical levels*, when we can put the "feel sorry for" aside just long enough to realize that, "hey, maybe its not my fault that so-and-so, which I had not seen in a decade, suddenly decided to jump off a bridge" and so forth. In the light, all of these types of *Imprints* really *do* seem rather silly and illogical—but we carry them anyways, up until the point that we can really examine, analyze and evaluate them in the light of *Awareness*. What other appropriate definition for "true enlightenment" could there be?

At a fully actualized level of I-AM (7.0), the individuated *Alpha Spirit*—"I"/*Self*—is only interested in being the *creative source* and *cause* of anything it is to direct "to be" in existence—and this provides a certain degree of fluidity to the Spiritual Universe (AN). Once we move to the area of the ZU-line and *Standard Model* that interacts with other *beta-existences* and *beta-lifeforms*, we find that suddenly there are many "*beta personalities*" running around this Physical Universe (KI). We have a *realization* that we are sharing *this* Universal Reality with other "MCCs"—each being directed by *someone* "to be" the *cause* in the Physical Universe (KI). This is a fact. Everyone want to be a *cause*. Even when someone is attempting to *assist you* in some way, they are doing it to be a *cause*, to extend their reach beyond *Self*.

This subject matter of "Help" is actually very closely tied to energies associated with those solids resulting from the *"blame-shame-game"*—meaning: times when you have rejected someone else's assistance/help and times when your own efforts to direct assistance/help were rejected. These should be *Processed* too. And so long as you are able to confront them with high-frequency energy, there is no reason—for what is described in *this* chapter-lesson—to have to resort to the more *intensive cathartic* methods demonstrated in a previous volume as "RR-SP-1."‡ True realization is more powerful than what most discover naturally by observation through cloudy lenses of fragmentation. We all carry *analytical* abilities, but are not necessarily encouraged to develop or use them in this society which is authoritarian based and expects its population to simply accept and agree blindly.

SELF-DIRECTING CAUSE FOR EMOTIONAL EFFECT

The systems present in our society are designed to weigh down and restrict the Human Condition. They were designed that way—probably were once quite efficient during ancient primitive times when humans were little more than slaves (*"effects"*) to a higher directing power. Of course, not much has changed. There are more systems in place in between, which disguise the *effort* placed on us—but the *effect* is undeniably the same. The more *solidification* and *fragmentation,* the more complex the systems must become to remain fluid in their appearance of "variety" because so much has become *fixed*. This is directly why our modern society has become so convoluted. Too many *creations* with unassigned *ownership*. All of the materials, elements, minerals, essences and waters on the planet have always been here—and they have been reshaped and reformed many times. Curiously, modern Human society is in the habit of creating without responsibility—and now we have more masses standing in our way than we know

---

‡ See extended course section in *"Tablets of Destiny."*

what to do with. At least this is how things are observed in appearance—and that is what's important when we are dealing with the subject of Reality—the appearance of Reality and the Reality of appearances.

When an individual makes someone else a *cause* through "blame" and "misappropriated responsibility" they are lending to them the power of *Self-direction*. As this develops over time—as it is reinforced repeatedly through practice, affirmation and/or behavior—these aspects become emotionally encoded as an *Imprint*. This type of *fragmentation* represents many common reasons that Human Potential is never fully actualized in modern society. The more we consciously make others responsible for what happens to us, the less that we are able to manage what is actually within our responsibility. Likewise: the more we consciously accept inappropriate blame and guilt, the greater our emotional *imprinting*—but that can be *processed* very easily with AR-SP-2. And after elevating *Awareness*, a person operating in high-frequency can sustain their vibrations in the face of simple communication about blame without actually succumbing to the lower-level emotional states. In fact, a fully actualized WILL should be able to sustain confrontation of lower *beta* energies without succumbing to them. That is certainly one of the goals in our *Systemology* work.

In most instances where an individual has taken on the *beta-experience* of an "ill effect," there is a tendency to quickly assign "blame" and "responsibility" to an external source. This is a "natural reaction" of *beta-systems* of *Awareness*: to want to *distance* the *Self* as far as possible from the "ill effects" as a primitive survival mechanism—but it is not *Self-Honest* and does not promote *Self* in actualizing *Self*, because it is, again, being removed from the equation as cause. It becomes clear after even some light education regarding *NexGen Systemology* that: the 21st century is wrought with an infinite amount of ways to remove, dampen and lessen *Awareness*, leaving very few valid methods to actually regain, maintain or even become *Awakened* to latent potential of the Human Condition.

Drawing upon our own research into the tendencies and similarities between early *cases*—studied as a means to continuously improve and advance *Systemology* work—there is one common occurrence observed regarding the Human Condition that may actually fall within a conjoined responsibility with those we interact with.

I am referring to the old game of: "*You made me feel ___.*"

How many times have we been presented with information suggesting we are responsible for how *others* feel and what *others* decided to do and so forth. This is directly relevant to our subject of WILL. An individual actualizing personal *Will-Intention* in high-frequency *beta-existence* will demonstrate recognizably higher degrees of "personal magnetism" and "charisma" in daily life, thereby essentially fending off the *effects* of lower-level vibrations. The *Seeker* should be consistently working toward a state of *impervious composure*.

—RECALL a moment when you heard someone say the phrase: *You're bringing me down.*

—RECALL a moment when you said the phrase: *You're bringing me down.*

—RECALL a time you saw someone entering a room.

—RECALL an instance when you *felt* someone enter a room.

—RECALL a time you saw someone exiting a room.

—RECALL an instance when you *felt* someone leave a room.

—RECALL a time when you *made* someone cry.

—RECALL a time when you were *made* to cry by someone else.

—RECALL an event when you watched someone *make* someone else cry.

—RECALL a moment when you tried to *make* someone feel better and were rejected.

—RECALL a moment when someone tried to *make* you feel better and you rejected them.

—RECALL an event when you watched someone try to *make* someone else feel better and they were rejected.

—RECALL a moment when you tried to *make* someone feel better and succeeded.

—RECALL a moment when someone tried to make you feel better and they succeeded.

—RECALL an event when you watched someone *make* someone else feel better and they succeeded.

—IMAGINE your physical body is enshrouded in a *sphere of light*.

The fact that emotional energy is carried and communicated by all individuals should not be a surprise. What is interesting, however, is that due to the low-frequency of ZU by which emotions are exhibited, the responsibility for them is rated to similar degrees as *physical* actions. An individual can direct higher-frequency harmonics of "emotional energy" after elevating *Awareness* to the aesthetic range of WILL (5.0)—thereby directing expressions of true creativity and imagination. This is a higher level of communication, exceeding the band of *beta-thought*. This is why there is so much "intrigue" tied up in aesthetics and art—and why such aspects are considered "beyond reason and intellect" when compared to "knowledge" in *beta-existence*. This is where the abilities of *Self* often seem to appear quite "magical"—because instead of remaining solely reactive to our environment, or fixed to programming and *beta-thought* that is communicable in words and language, the *Self*, at the actualized level of WILL (5.0) and above, has abilities to *intend* and *Self-direct intention* towards its own emotional manifestation in *beta-existence*. This ZU would normally be channeled into *solidifying Imprints*, but it can also be *Self-directed* to intentionally "charge"[135]

---

[135] **charge** : to fill or furnish with a quality; to supply with energy; to lay a command upon; in *NexGen Systemology*—to imbue with intention; to overspread with emotion; application of *Self-directed* "intention"

ourselves, our environment and/or other objects, with any *beta-energy* at *Will*.

Although we have used the term now a dozen times for introductory semantic convenience, it should be understood that (with the exception of a few examples of physical enforcement) no individual can *make* another individual *feel* anything —what we do is communicate our *Will* in such a way that *conjures* or *invokes* the "inspiration" for a *feeling* or *emotion*. An individual must either be in a predisposed state of agreements with a feeling that is now validated, or they are maintaining a low-frequency ZU state that simply puts them in a position to be more suggestible at that moment. Our basic energetic state is always communicated and others are generally able to sense that, even when they are not absolutely certain what the source of that perception really is. However, this can also form into *fragmented imprints* if treated as such over time—whereby a certain person, place or *facet* is now linked to a certain type of experienced effect. This borders on the "psychosomatic"—whereby the unprocessed emotional *Imprinting* we reinforce is able to actually *cause us* to *feel* a certain way due to reactive-response encoding. We can actually make ourselves violently sick and even create disease just to validate our beliefs about the effects that some external presence has over on us as cause. This is a surefire way to diminish personal WILL.

The power of WILL to *Self-direct* an intended emotional wave of energy in *beta-existence* is a common practice (or intention) of nearly all spiritual, mystical, religious, and New Age methods, traditions, techniques, prayers, rituals and creative visualizations. There are far too many names and variations of these to be worthwhile for this present volume—however, suffice it to say that *there is a reason* that initiates, practitioners and congregations around the world, since the inception of the current Human civilization, have discovered *any* degree of effectiveness in such methods. The one key feature

---

(Will) toward an emotional manifestation in beta-existence.

that is common to them all is: the realization of WILL and actualization of *Intention*.

Emotional energy can be created by WILL (*Intention*) without succumbing to the actual low-level *emotional fragmentation* in-and-of itself. As *cause*, WILL intends *effect*. It does not require exciting a personal display of *emotion* to accomplish this either—it may be accomplished solely due to *Alpha Intention*. This is precisely how the "I AM" as *Alpha Spirit* directs all *cause* and *consciousness activity* form the "ACC" (7.0). Everything above (4.0) on the *Standard Model* and ZU-line (and any *Systemology* model, chart or scale) is considered "causal" in terms of *beta-existence*. Using the power of *Intention*, the *Seeker* can simply "*Will*" a desired *effect* to take place "lower" down along the ZU-line.

As long as a state or condition is *Self-directed*, created by *Self* and is "owned" as a responsibility of personal power, there is no danger of lasting *fragmentation* in *willing* any variety of activities along the ZU-line. So long as a *Seeker* is fully *Aware* that *imprints* and *beliefs* are their own creations to form and transform—and also recall energy from during *dissolution*—than any constructive uses of WILL are limited solely to one's own imagination and creative expression—which are *Alpha* qualities.

The *Seeker* might practice this energetic work using more esoteric methods derived from the ancient mystical schools of philosophy. An initiate was told to take an "ordinary" object, such as a rock/stone, or small piece of metal—some kind of trinket—and "charge" it with an "intention." Although there are many ritualistic and dramatic techniques recorded throughout many old dusty volumes regarding this, the basic principle of them all is always the same: *Self-direction* of *Intention* via *Will*. All that is required, is to:

> bring up, conjure, recall or imagine a certain aspect to the fullest extent of "*Alpha Thought*" that you are able—directing the most complete and detailed impression of

the "archetype" you wish to create and then releasing it toward the object with the fullest extent of *Will-Intention*.

The concept is actually rather simple—however, its mastery was very highly prized among ancient shamans, magicians and wizards, and so a great deal of time and lore is extended toward the various methods and types of practice that accomplishes these *effects*. But, they are nearly all variations of the same basic premise, which may be operated on any neutral object in your vicinity as practice.

OBJECTIVE PROCESSING
(SELF-DIRECTING WILL)

PART A

—IMAGINE your physical body is enshrouded in a *sphere of light*.

—LOOK around and IDENTIFY a small neutral object you could hold.

—DECIDE that *you will* go and pick up the object; and then *you* do it.

—IMAGINE the object is a sentient being that is capable of emotion.

—CREATE the emotional feeling of *happiness/joy* and hold it in Mind.

—WILL the emotional feeling of *happiness/joy* into Being.

—INTEND the object to fully experience *happiness/joy*; and then Will it.

—REPEAT the intention and direction of Will several times until *you* are *certain* that the object is experiencing and expressing *happiness/joy*.

—REPEAT the above four steps using the emotional feeling of *sadness/grief*.

—REPEAT the primary steps using the emotional feeling of *ferocity/anger*.

—REPEAT the primary steps completely, alternating emotional feelings fully, making certain that your own emotional state remains independent of emotional fluctuation. Practice repeatedly. (*End session on a neutral emotion, just as at the start.*)

PART B*

—INTEND the object to be *fiercely angry*.

—RESURFACE an emotional feeling of *fear* of the object. ANALYZE any *facets* you associate with this feeling.

—IMAGINE the object is genuinely *afraid* of you.

—RESURFACE an emotional feeling of *anger* toward the object. ANALYZE any *facets* you associate with this feeling.

—INTEND the object to be experiencing *grief-stricken sadness*.

—RESURFACE an emotional feeling of *sympathy* for the object. ANALYZE any *facets* you associate with this feeling.

—IMAGINE the object to be *sympathetic* toward you.

—RESURFACE an emotional feeling of *relief* and *interest*. ANALYZE any *facets* you associate with this feeling.

—INTEND the object to be experiencing *happiness* and *joy*.

—RESURFACE an emotional feeling of *happiness* and *joy*. ANALYZE any *facets* you associate with this feeling.

—REPEAT the above steps (of PART B) completely several times until you start to feel that you have greater personal management (*Self-control*) of directed emotional energy and the response-reactions tied to emotional energy from external sources.

---

* Part-B may be practiced as an extension of the previous Part-A session. If practiced as a separate session, use the first four steps from Part-A to properly establish the *Process*.

OBJECTIVE PROCESSING—"PART C"

Many variations and applications of the above exercises may be derived. The goal of this *Processing* has been achieved (for the present cycle of development) when the *Seeker* has reached a point of personal certainty (knowing) that a Willed *Intention* is carried to action. Ongoing purposes for *objective processing* always involve increasing the *Seeker's* certainty regarding personal abilities to manage *Self-direction*: in this case, the very literal *Will-Intention*, an "emotional equivalent" operated by *Alpha Thought*—"directions" from *Self* that are *Willed* to *beta-Awareness* by *Intention*. And thereafter, the entire orchestra[136] of personal interaction ensues from the MCC (at 4.0) all the way down the ZU-line to produce a causal effect in the Physical Universe (KI). When these *effects* are not realized in the Physical Universe (KI) or *beta-existence*, the emotional energy has a tendency to wind up, collect or solidify as an *Imprint-Image*, which later distorts a *Seeker's* clear view of the Physical (1.0) or continuity of KI (0.0) from the *Light of Awareness* that is shined down from (7.0) the *Alpha Spirit*—the true position of *Self*.

The full *Self-direction* of WILL (5.0) is rooted in *Self*-control of "*Alpha Thought*" (6.0). Combined, these *Alpha* systems result in the "force" or "pressure" of personal *Intention* that imposes onto *beta-existence* all along down the ZU-line of energetic interaction. In fact, the very definition of "magic" composed by magicians of the early 20th Century was: "to *effect* a *change* in *Reality* in accordance with one's *Will*." Well, as we have demonstrated throughout the cycle of work in the present volume, "effecting a change in Reality" is the very *Game* that the *Alpha Spirit* is always playing with—but getting *others* to *agree* to that Reality; ...now, that is the trick.

To be of the utmost highest optimum effectiveness requires WILL to be *Self-directed* with the absolute totality of certainty available as *Self*. The *Alpha Self* is spiritually free at (7.0) "to

---

136 **orchestration** : to arrange or compose the performance of a system.

be" anything and conceive of anything—just as we have a greater range of activity in the beta Mind-System than we do with physical activity. The initial qualities "to be" are only tempered or filtered by other *Alpha Systems* when they are "to be" *Willed* into *beta-existence*. This is in the same basic function of systems that we find concerning *beta-direction* of the MCC and the energy working its way through the range of "thought channels" before brought into "physical effect" by "physical effort." This exact same type of sequencing is taking place at *Alpha* levels as well. But initially, a *Seeker* is best equated with a realization of these higher systems by directly relating them to their counterparts as *beta-systems*—hence the emphasis found in the present volume and its former —"*Tablets of Destiny*" materials—on an educational level. *Grade-III Processing* is generally directed toward *beta-fragmentation* and *Imprints* accrued during *this* lifetime—unless something *otherwise* directly resurfaces on its own. *Standard Processing (SP)* Techniques may just as easily apply to these *facets* as well—it simply is not our present focus.

Additional *objective processing* may be easily developed from applications of esoteric energy-work and mystical training passed down to us from the Ancient Mystery School. These initiates would dedicate many years toward a disciplined development of what many today might consider very basic and/or trivial exercises. Yet, some of these very obscure—but simple and workable—methods were found to be highly effective in elevating consciousness when treated over extended periods of personal development. We discovered many of these techniques were first revisited during our modern age most vigorously in Western Europe during the late 1800's—when these ancient methodologies were revived in public sight by "mystical orders," "esoteric philosophy schools" and other exclusive intellectual "societies" that gathered, educated and practiced privately by INITIATION. In America, these same traditions were best realized by an underground—though publicly visible—intellectual movement known as the "New Thought." In many ways, *NexGen Systemo-*

logy could be considered a futurist extension of both these previous modern *realizations*, as a means of *actualizing* the best future possible; taking only the best processes from what is even remotely workable from the past for our purposes.

> PART C is a basic demonstration of the axiom that: "All Intentional Acts are Magical Acts" (as stated to us by our esoteric predecessors). This process is similar to *earlier* parts except that the goal is to *realize* a direct communication line of *Will-Intention* between the *Seeker* and the *object*. This includes communication of *Will-Intention* as "control" (using *command lines*) and an "acknowledgment," regarding the *object* as endowed with sentient faculties you can contact and *with* motor functions. (The *object* is treated as a living being.) The *Seeker* makes a *Self-directed* command of *Will-Intention* to the *object* to the fullest extent imaginable. Then the *Seeker* actually performs the physical action to achieve the desired *effect* and genuinely "thanks" the *object* as to acknowledge the *result*.

In former *Grade I* materials—used, for example, in a "magical school"—an initiate might be given armfuls of fancy tomes relaying all manner of quasi-philosophical rituals and bizarre ceremonial activities to conduct involving this or that type of herb, candle, altar attire and so forth; all of which are *tools* meant to redirect the *Awareness* of *Self* onto the simple fact that everything we do and focus on to the extent of our *Attention* and WILL becomes "magical." And what is that? What is something "magical"? It is imbued with a quality that we have not otherwise been able to pinpoint semantically with *beta-sciences*—and that missing key *is* the matter of ZU, and the "attention-units" of *consciousness*, and the power of the Observer, &tc.

The most widely used and basic example of PART C is found in many of the "metapsychology-parapsychology-psychic" type research organizations, schools and literature regarding the subject of "telekinesis"—or else the mental command of objects to move. This is not necessarily the purpose of presenting this *objective processing*—but, perhaps at another level, it is not excluded. It may be that such effects are *actualized* completely only after the present Human Condition has fully attained its spiritual evolution into *Homo Novus*. The reason this is mentioned at all in present context is not to *bait* the *Seeker* further on a *Path* toward evolution, but simply to point out a direct and verifiable fact that these developmental arcane techniques—in all relevant esoteric sources—were found to deliver this same basic methodology to their initiates, for whatever *their* own purposes intended.

There are really only a handful of valid esoteric points to consider regarding "PART C." A *Seeker* should first make certain the object *is* real—the first step of *realization*—which could be a matter of *Self-directed* "Identification" (covered in previous *Processes*). There are also *facets* associated with the *object* as a Reality—meaning anything you may *like* or *dislike* about it. To be effective, a *Seeker* must essentially find it acceptable *to be* that *object*, even if for a moment. So there must be some degree of shared *affinity*[137] and *agreement* on the "Reality" shared between *subject* and *object* in order to have valid and effective "communication"—which is any direct transmission, activity or motion of energy, information, data, &tc.

> All "communication" (interaction) is clear and direct
> from its origination as intended with certainty
> up to its degree of *fragmentation* at any relay point.

It is important too that a *Seeker* does not get "too caught up" in this exercise. It merely stands out—as do many of the more *objective processes*—because of the physical activity involved and its similarity to other mystical training techniques. It is

---

137 **affinity** : the apparent and energetic *relationship* between substances or bodies; the degree of *attraction* or repulsion between things based on natural forces; the *similitude* of frequencies or waveforms; the degree of *interconnection* between systems.

no more or less significant or important than other practices used to *actualize* and *Self-direct* the WILL—but its greater applications in "other areas" are unmistakable and easy to practice *without* first involving "another individual." The fundamental principle is that an individual is directing WILL with the certainty that the effect will be carried out as intended. Traditionally in *NexGen Systemology*: it is run on a *Seeker* by a *Pilot* to the point where the individual is no longer surprised or exhibiting any reaction that the *object* is not moving (obeying) on its own motors.

From the several dozens of examples studied for intellectual validation on PART C, it is interesting to note that nearly all (except some of the most recent publications) seemed to prefer using a *plain clean circular ashtray* as the focal object—preferably one that is not transparent glass. However, in even the cases where it was not an ashtray it was usually some type of "container" that *Will-Intention* could be put "into"—so as to fill it up with an intention or command. All of the example items suggested could also be charged with a basic movement (and command) such as "stay there," "stand up," "sit down,"—in the same way we would *direct* any other life form we took responsibility for, but with the fullest extent of our *Will-Intention*. These actions are also easy for us to *affect* when we physically make the change and acknowledge it. By repeatedly doing so, we achieve a greater pattern of certainty in our "surface thoughts" to create change.

The basic methodology presented in this unit of the present volume carries a very real solution when developed appropriately. An individual with difficulties *"recalling"* moments demonstrating successes from efforts in the past will demonstrate greater difficulties later on when expressing *Self-direction* with any certainty. An individual used to being "held back" gets into a pattern that the walls and barriers are there restricting them even when later they are not. These same conditioned patterns may be re-conditioned. In the end, all of these responsibilities for *Self* will fall back onto the *Seeker*—

the individual themselves—to manage. You might as well start right now. What you carry is yours to recognize, own and dissolve as you *Will*—but none of it may be ignored if your personal goal is to *reach the top*.

SELF-DIRECTING WILL IN SELF-HONESTY

"Live neither in the past, present, nor the future—
but in the eternal."*

The fundamentals of *NexGen Systemology* are grounded on foundations that extend back on the timeline of human history to the inception of the current "systems" inherent in our modern world. These records—mostly maintained on *cuneiform tablets* from Mesopotamia—are very often referred to as the *"Arcane Tablets"* within *Systemology* literature and its predecessor, *Mardukite Zuism*. Although many interpretations, analytical philosophies and academic approaches have already applied their treatments to the remnants of the Ancient Mystery School to the extent of their own selected paradigms, the fact remains that the deepest meanings *realized* on the *"Arcane Tablets"*—from thousands and thousands of years ago—is only *now* being *actualized* toward a full application as a futurist spiritual technology for the first time in modern history. And yet, we are only now *discovering* these new uncharted territories, which have *always* been there for us. And as we travel that *Pathway* we are in constant reminder that many elements on this journey are *all too familiar*—and *realize* that we *have* actually charted this adventure once before. But, of course we have—for this is, and always has been, the *Adventure of Self*.

When the *Seeker* has reached an attainment of actualized *Awareness* outside of *beta-existence*, therein alone lies the true personality of the *Alpha Spirit*, the individuated "I" that is *Self*. This is the state of *Self* at (5.0) free of *beta-fragmentation*. The

---

* Quoting the *Theosopher*, Mabel Collins in *"Light on the Path."*

Seeker has extended the reach of *Awareness* and the ability (and responsibility) of *Self-direction* from the point of WILL (5.0) as *Cause* in *beta-existence*. The *Seeker* has there risen above the planes of *Effect* and *Desire* as they pertain to the Physical Universe (KI) beneath "Cosmic Law" of causality. At (5.0) the *Seeker* is still an individuated *Alpha Spirit* beneath the ALL—the "LAW of the Spiritual Universe" (AN)—but they are now able to more clearly enact *Self-directed effects* in *beta-existence* that are free from lower levels of personal *fragmentation* and *programming*. They are applying *Will-Intention* to the fullest extent from outside the "beta-systems" without regard of former *fragmentation* and erroneous *programming* concerning personal ability and responsibility accumulated during *beta-existence*. The way out is always the way through; not avoidance, neglect or indifference—which mark a path to ignorance.

Without going into a whole semantic philosophy lesson, we have—above—brought up a new word that has not really been applied to *Systemology* vocabulary and that is *Desire*. It should be understood that while *Awareness* remains in *beta*-states occupied within the Physical Universe *from the inside*, the individual always remains—to *some* significant degree—in the domain of *effect*. Such is simply an existential by-product of being fixed to the Human Condition. But—rather than assign *Desire* to the *ZU-line* or systematic Scale-Charts with a particular value, we tend to treat *Desire* along the same lines as we do *Effect*, at one direction of the spectrum—with *Cause* and *Will* at the other.

When we are not describing intellectual states of "interest" and "determination," the real perception of *Desire* is actually a *facet* of *fragmentation*. This is probably one of the reasons why the Buddhists don't want anything to do with it. An individual operating from a point of *Hope* and *Desire* is still attempting to resolve the need to "*have*" something—and so they are "yearning" to *Will*, by exercising *Hope* and *Desire*—they are still "desiring *to be* WILL."

For many individuals—this is the extent of their experience of the Human Condition.

In the ancient cosmological tablet narrative of the *Enuma Eliš* —the Babylonian "Epic of Creation"—given formerly in the *"Tablets of Destiny"* volume: the Anunnaki figure MARDUK is given the *"Arcane Tablets"* and with this knowledge, he *realizes* the ALL and *actualizes* his WILL in accordance with the LAW. As a result of one supreme execution of "Alpha Thought" (6.0) via WILL (5.0), he defined and structured all systems of the Physical Universe (KI) under "Cosmic Law"— also called "cosmic ordering" in *NexGen Systemology*. This may well be a mythographic model—in fact, other *tablet* sources suggest the Babylonian reassignment of this WILL to MARDUK in order to establish "Mardukite Babylon" and the world's first systematized religious tradition. But, that only poses questions as to the identity of a "creator" and not the manner of the "creation" that is rendered in arcane literature, when reading *between* the lines—which contributed extensively to our *Standard Model*.

As a *Seeker* makes their way along the *Pathway to Self-Honesty* the direction is always forward toward a *unity with Self*—the combined and actualized *Awareness* as *Self*, looking outward and down the entire length of the ZU-line without *fragmentation*. Certainly the full scope of this cannot be guaranteed by simple delivery of a singular standardized demonstration to the extent of this one book—even one of such prestigious magnitude as the present volume. However, we are not slighting out any potential effects or limiting possible results. It is obvious that although we all share similar qualities and abilities as *Alpha Spirits*, each individual has traveled their own *downward spiral of separations* and *fragmentation* from *Self* —and the true original total actualized *Awareness* for-and-as *Self*. This is one reason why the journey back is so familiar and also is very much individualized to the route that you already have taken to arrive where you are.

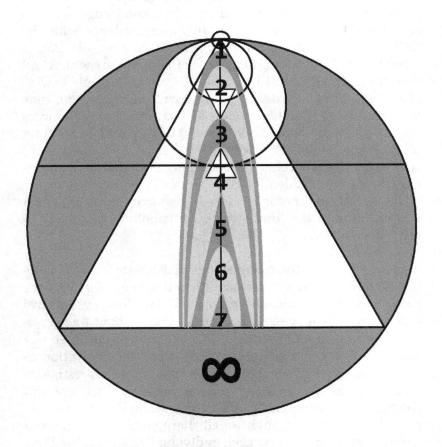

SYSTEMOLOGY PROCESS "SP-2B-8A"[‡]

The following esoteric exercise is derived from the "*Arcane Tablets*" as a practice toward *realization* of merging "*individual consciousness*" with "*Cosmic Consciousness*," which is to say the highest state of being as an Ocean of Infinity (8.0)—written as AB.ZU in *Sumerian* cuneiform.[∞] Even as a practice, such *realization processing* enables the *Seeker* to more easily *realize*—and eventually *actualize*—the highest state of Cosmic Will, as individual potential increases with the unfoldment of *Self-Actualization*. The purpose of such "highest level" exercises in mystical schools and spiritual traditions is not necessarily *to be* the Ocean of Infinity (8.0), but to actualize *Self* (7.0) as a *Total Awareness* of the *Alpha Spirit* "I" amidst a focal center or wave-peak of that Ocean. To fully actualize this from *beta-existence*, one would necessarily have to clear *ZU-channels* of *Awareness* up unto the point of (7.0)—however, the bulk of mystical teachings are found to contain very specific focus-directing creative exercises that often seem trivial at face value but of which actually are demonstrable to achieve effective results when conducted as a continuous regimen of personal practice over time.

—IMAGINE your physical body is enshrouded in a *sphere of light*.

—FOCUS your *Awareness* on the *Eighth Sphere* of *Infinity*.

—IMAGINE the *Infinity* of *Nothingness* extending out "infinitely" on all sides as a great Ocean of Cosmic Consciousness.

—FOCUS your *Awareness* from *Self* as a singular focal point of individuated consciousness in the center of the *Infinite Ocean*.

[‡] Supplementing the "SP-2B-8" Self-realization process using the "*8 Spheres & Circles*" from the *Halloween-2019 Extended Course Lecture* (given previously in this present volume).

[∞] Spoken and transliterated as "AB-ZU" or "*Abzu*"—although actual written tablet renderings use the cuneiform signs ZU+AB for the expression.

—SENSE that the *Nothingness-Space* all around you is rising up as tides and wave-actions of invisible motion; its abyssal stillness broken by the singular point that is *You*.

—SENSE that as you press your *Awareness* against the *Nothingness*, there is no resistance, there is no sensation—no feeling of any kind.

—IMAGINE your totality of *Awareness* as the singular focal point of *Infinity*—then REALIZE that the waves you see crashing up against you and rippling into *Infinity* are an extension of your every thought, will and action.

—REALIZE that you are the *Alpha Spirit*; that "wave peak" in an otherwise *Infinity of Nothingness* stretching out within and back of all that was, is and ever WILL.

—REALIZE that your conscious *Awareness* as "I", your direction of WILL as *Alpha Spirit*, and the "central wave peak" born out of *Infinity* are all the same pure individuated ZU— are all *One; None; Infinity*.

—WILL yourself to project *Awareness* ahead of you and see an extension of this ZU as your projection of Identity extending infinitely in front of you—all the way to the *zero-point-continuity* of existence—and back to *Infinity*.

—REPEAT this several times, IMAGINING this ZU as a *Clear Light* radiant extension from *Self*, directed across *Infinity* to *Zero-point* and back to *Infinity*; then REALIZE that you are dissolving and wiping out all *fragmentation* from the channel as you direct the *Clear Light*.

—REPEAT this several times, until you feel confidant in your current results for this cycle of work.

—RECALL the moment you last imagined your physical body enshrouded in a *sphere of light*.

—RECALL the instance you decided to start this present *session*—get a sense of the Intention you *Willed* to begin the session.

—REALIZE that your *beta-Awareness* and the true WILL of the Alpha Spirit are One; End the session.

SELF-DIRECTING MENTAL IMAGERY

The *Reactive Control Center* (RCC) of an individual's physical existence is closely tied to emotions and emotional responses—as we have demonstrated. But, what about our thoughts? What about the pictures that are inspired by—and which have the ability to inspire—other emotions? What about the images in our mind that we associate with emotions, events and yes, even basic "words"? We have all heard the old expression about *not* thinking of an *elephant*; but then you can't help thinking of the *elephant* and now you have an *elephant* in the room with you. Actualizing inherent ability of *Self-directed* WILL (5.0) necessarily depends on personal control and management of *beta-Thought*—which is the conduit or channel of communication on the ZU-line between the RCC (2.0) and MCC (4.0). "*Alpha Defragmentation*" of WILL runs between (4.1) and (5.0) and is mostly concerned with freeing up and disinhibiting personal *creative* expression and *imagination*—essentially extending the "reach" of the Mind-Systems during *beta-existence*. This constitutes what some have described as "transcendental experiences."

Creative expression—the use of symbols and abstraction—is not a *beta* quality and may only be assigned to *"spiritual consciousness"* or else, the *Alpha-Spirit*. Beyond *beta-existence* of the Physical Universe (KI) we discover all of the individual *causes*—games; logics; universes; Alpha Spirits directing Alpha Thought... This is where the form of "true archetypes" are cultivated, which are materializing in *"mental consciousness."* This memory and knowledge extends far beyond anything we have directly experienced in this lifetime—or anything programmed at genetic levels of the *biological organism*. Much of what is stored for creativity actually comes from a higher spiritual version of *Imprinting* that contains our experiences for the entire spiritual existence of *Self* independent of the genetic memory of the *genetic vehicle* now being operated. This is the energy or ZU-frequency "band" that is responsible for development (and high-level appreciation) of innovation, art, music, verse... all of the aesthetics behind any sense or experience *realized*.

> If the Pathway to Self-Honesty provides anything for the average reader, it is simply a demonstration of the very *automated, reactive-response, conditional* and *controllable* natures of diverse Systems operating between an experience of "I-AM" (7.0) and the energetic-material continuity of the Physical Universe (0.0).

External efforts (counter-efforts from others) to control your WILL—and inhibit your creative ability and personal exploration of imagination—are demonstrations of *enforced reality*. Others are enforcing their WILL in order to exercise *cause* and make you an *effect*. We all develop thought-forms based on our *Will-Intention* and other *beliefs*; we put them into action and we *watch*. We change a variable in the system, or we don't, but *we decide* it. All existences *"exist"* for the sole purpose *to be* an existence. The manner of your own *existence* is decided purely for and by you—the continuation of that existence is based fundamentally on your ability to *decide correctly*.

Examine the idea of *"mastery* of any *skill"* and you will see a few common factors there: *practice* is one; *repetition* is another; there is an intriguing idea of *muscle memory*; but none of the factors are found within the realm of *intellect*. A person truly *"skilled"* in his art or craft does so by WILL and *Intention*. When asked how they do it, the *Self-Honest* response is always the same: *"I just do."* There is no *belief* or *Alpha Thought* in place to *Self-determine* otherwise.

While studying this material—and the former volume *"Tablets of Destiny"*—a *Seeker* may already have reached a conclusion that automation of systems simply *is*, and as much as it may be used to condition us, these same techniques may be *Self-directed* in order to "un-condition" our *Self*. Since we have demonstrated that *Self* is never "less than" by *shedding skin* of fragmentation, there should be no fear or reservations about progressing further upon the *Pathway*. At this stage of *Self-Processing*, a *Seeker* that is unable to "let go" is still requiring resolution to *"havingness"* in *beta-existence*. Such an individual is still in a *set* where they have *postulated* that they will be "losing something" by dissolving layers of artificial material formed around the *Self*.

> A close examination of how a person maintains, purchases, organizes and manages "physical possessions" will also reflect how they treat "things" formed in the Mind-Systems—because to *Self*, at any level of *Awareness*, both are *just* "things."

Self-Processing directly excites, practices and repeats methods of *Self-directed* "creativeness." This, in-and-of itself builds an amount of certainty in regard to *Self*—which is why there is some effectiveness to the former "motivational guru"/ "change your mind; change your life" type of behavioral modification approaches popularized throughout the late 20th Century. These methods, from the point-of-view of *NexGen Systemology*, have all been discovered to be simply "pop-culture" demonstrations of very basic "New Thought" principles. Then they combined this with the same rudimentary "behavioral-brain-psychology" that we are now pushing to get away from with our new *NexGen* 21st Century *realization* of a *new* "New Thought" as *Systemology*—an "applied spiritual tech-philosophy" that is *actually* workable and effective for the 21st Century *"spiritual transhumanist."* There has been no other valid representation of this in modern times simply because it is not as easily "marketed and profited" from—which is the extent low-level *actualization* gets a person in this world.

> Now *Seekers* have two new modes to consider for the 21st Century: *Mardukite Zuism* and *NexGen Systemology*—which, of course are not mutually exclusive.

We have now realized that every technique and method of *Systemology Processing* that targets personal management of emotion and memory assists *Seekers* in regaining *Self-control* of these states (systems). This is practiced repeatedly, because we are dealing with *conditioning*. And we are not using *Systemology* to condition ourselves into some false sense of grandeur. What we *are* doing is systematically removing *fragmentation* that undeniable reveals the "*Unconditioned Self*"—the very *Alpha* state of our true spiritual *Beingness*. There is no higher purpose or mystical demonstration of the "Great Work" available in *beta-existence*.

Virtually every other cultural mythography and spiritual paradigm has attempted *to be* the route toward *Ascension*—and perhaps within each parameter of the System across time and geography, those methods and ideals were executed to some degree of *effectiveness*. But, we are in the 21st Century now, beginning a decade of 2020's—and it is time we brought some *clear 20/20 vision* to the "I" that is *Self*, and actualize a new *realization* of the Human Condition that unfolds the latent *genetic and spiritual* potential that is already present and just waiting to advance us toward Homo Novus in *this* lifetime! This is *the most important* thing an individual can be *doing right now* for themselves, for family, for society, for all *Life* on the planet—and to every sphere of influence and existence that we can extend our reach in *Self-Honesty*.

Without complicating *defragmentation procedures* with a series of superfluous stepped processes, it should suffice to explain how a *Seeker* would work to further develop *visualization* skills —which includes:

a) the ability to WILL *holographic imagery* out into the Mind's Eye (or Third Eye); and

b) the ability to manipulate/change and dissolve/destroy these "illusions" at *Will*.

So long as this condition is always met, there is no *fragmentation* by our own "illusions" as long as we *know* they are our own "illusions."

When we start to pass *cause* over to some other source or object or *Identity*, then we start to run into personal turbulence—and a reduction in "power" via the reduction in responsibility and the mis-assignment of the *Creator* as the *source* of Reality.

[This perfectly describes what happens regarding *fragmentation* in more ways than will be readily apparent at first, so you are encouraged to read this particular statement several times before continuing on.]

Whenever an individual says a word, or you think of a word, there are associations made in the Mind-System *automatically*. Whenever we see something or hear something in our environment, we tend to carry some resonance of it around with us as part of our memory and experience. For example: as a young child you may have seen that adults were using certain things to get around in. They may have come in many shapes and sizes, without clear definition, so they blurred together as "these big things that sometimes are still and sometimes move and sometimes you can ride in them and—Oh, be careful now because don't get too close to one and such" all becomes a series of associations that eventually are imprinted intellectually and experientially complete with emotional reinforcement with the word: "CAR" as a *Reality*.

Now, perhaps at first, before even more recent times of technological development, there was no distinction of CAR, for example, from AUTOMOBILE. Therefore, by invention (or systematization) of "motor-vehicles" and the former concept of "automobile," suddenly there was an entirely new "thing" to have *associations with* and "know" *something about*. And these "facts" might all be demonstrated quite clearly to be "true." So, *now* someone can demonstrate to you that a "car" is not a "truck" (&tc.) and appear as though they have *something* to *know* about. Do you see? [These basic examples within this present volume should not be slighted out of importance due to their simplicity. They replace many chapter-lessons worth of demonstration that would otherwise continuously revolving around the same points—and undoubtedly drive a

Seeker into boredom or disinterest on these subjects.]

From the very beginning, this present volume was intended to cover a significant "amount of ground" for the *Seeker*. This has required us to move very swiftly through *realizations* and across material that in-and-of-itself could be used for direct attainment of the *Ultimate Goal*, without further assistance or material—*if* it were *actualized* fully. We are not slighting out any significant progress made on the *Pathway*—but we should not "stop" our motions forward each time we come to some new *realization* point, or sense of "arrival." Too often, the feeling comes over us at each critical point of *realization* that we "must be *there* now." Then, of course, in a second thought we realize, "well—maybe I could have applied a bit more WILL, or clearer energy, or not have had any reactive response at all to such and such, &tc." At each point we indeed *have* achieved something—and at each point we realize we *can do* just a little bit *better*, and get a little bit *further*, the next cycle around. And *that's* the *Game*.

> Systemology demonstrates that one way we may effectively *reduce fragmentation* and *restore* vitality of ZU-flow as *Awareness* is "exhaustion of thought"—which is also to say "exaggeration of thought."

An obsessive person may be running an "exaggerated expression" of thought in a loop when it is not *Self-directed*, but when we are actually putting conscious *Awareness* of the *Alpha-Spirit* into this repetition, something else happens entirely. We find a personal phenomenon that we now call "flattening a collapsed wave."[138] As explained in *"Tablets of Destiny"*—and to some degree in the general *quantum* inter-

---

138 **flattening a collapsed wave** or ***processing-out*** : to reduce *emotional encoding* of an *imprint* to zero; to dissolve a *wave-form* or *thought-formed* "solid" such as a "*belief*"; to completely run a *process* to its end, thereby *flattening* any previously "*collapsed-waves*" or *fragmentation* that is obstructing the *clear channel* of *Self-Awareness*; also referred to as "processing-out."

pretation of the Physical Universe (KI)—"*collapsing a wave*"[139] is: *Awareness* taking a wave of potential states and fixing it as a "peak" or specific arrangement of interaction. This seems, by description, much more *mystical* than it actually is. In essence, we take all that *could be* and make a decision as to what *is*; we are making a definitive decision to agree on what something is *to be* in exclusion to that it is *not to be*. It has now become an *effect*. It's solid and doesn't hold any more Light for us as we peer down from *Self*. If we "*flatten*" this obstruction—running it out in *processing* or another systematic treatment of *postulating*—than we are returning the "wave" to its original state of potentiality—thus we are able to use its power again or even *collapse the wave* as some other "thing" or "state" at WILL.

ALL PROCEEDS FROM AND THROUGH WILL

According to "*Arcane Tablets*": the *fragmented* Human Condition in *beta-existence* will always "act and decide" in accordance with personal programming (in the absence of true *Self-direction*), which modern society generally refers to as the "personality" or "character" of a person—and tries to demonstrate how these aspects are helpful "things" you *should* want. But, the *Seeker* on the *Pathway* knows better. When we consider the amount of time, attention and effort demonstrated in this present lifetime just to get *back up to zero*, it is clear that we should be working in the other direction, opposite the accumulation of more *programming*, erroneous solids, *imprinting* and layers of *enforced reality* keeping *Self* from experiencing *itself* as it should. We have clearly been *drugged, drunk, hypnotized* and *implanted* with all variety of emotions, thoughts, ideas, opinions and beliefs about our

---

139 **collapsing a wave** or **wave-function collapse** : the definition or calculation of a wave-function or interaction of potential interactions by Observation. The idea that the Observer is collapsing the wave-function by measuring it appears in the field of *Quantum Physics*. Consciousness or *Awareness* "collapses" the wave-function of energy and matter as the third (required) Principle of Apparent Manifestation.

Self—all of which only serve to keep us from a *Self-controlled* WILL.

Just as emotions may be *Self-directed* or *stimulated*, the WILL is called into action by energetic channels of "thought" or "consciousness" at levels immediately "above" and "below"—relatively speaking. At (5.0) the WILL might be engaged by either "Alpha Thought" (6.0 *and above*) or from *Beta*-Thought ("Mind-Systems")—and it is tied to emotional bands of *beta-existence* directly as EFFORT—which is simply application of *Will-Intention* gauged or filtered by the Mind. Even if you are still up and coming to understanding all the systematic valuations, components and steps regarding *Systemology*, there is at least one very clear demonstration point that you might take away from this and that is: regardless of all other activity described throughout this present volume, *Self* as "I AM" or "Alpha" at (7.0) is always *unchanged*. It may have a fragmented view; may witness scenes, images or ideas that are wrong and self-defeating (toward actualizing its own WILL), but it *is* wherever it *is*—and it remains whole.

There is a simple popular demonstration used during the last several hundred years whereby an instructor (or *Master*) would liken a pitcher of clear water to the "unconditioned Self" in its pure (Alpha) "spiritual" state. Then, poured out in clear crystal glasses, the *Self* could be demonstrate to "pass" clear light through it; the *Self* can see through it clearly; and to taste, it is pure and unadulterated. When we consider the influence of *emotion, conditioning, imprints* and the *reaction-responses* that "color" or "filter" our perceptions—this can be demonstrated by adding a food-grade coloring to each. Each demonstrates a different colored perception of emotion; altering the appearance and the type of light filtered through. But the composition of the water remains. We have also used various states of water to demonstrate other *Systemology* principles in the "*Tablets of Destiny*" material. The actualized WILL (of the water) *experiences* the color, but it does not *become* it.

We can control the nature of personal WILL and the *effort* produced from emotional energy on command. This simply requires the same type of practiced skill we might apply to commit any other knowledge or physical activity into memory. The *Pathway to Self-Honesty* provides an individual with increased knowledge, certainty and control of various aspects of the "mind" and "body" that otherwise run fairly automatically. When you close your eyes, what do you see? Is it energy patterns? Geometric shapes? Images from your lifetime or memory? Is it pure blackness? If there is mostly darkness, do you notice any variations? Practice taking a "snap-shot" image of whatever you might be seeing in your Mind's Eye—copying everything that you are able to perceive. WILL a copy of image to form next to the original image. Continue to do this several times. Try to stack the copies on top of each other. What does this do to the sensation and presence of this image? Does the image become more solid and dense or does it become more transparent and thin?

—IMAGINE your physical body is enshrouded in a *sphere of light*.

—LOOK around the room and IDENTIFY an *object* that you like.

—FOCUS your *attention* solely on the *object* and nothing else; no other external activity in the environment and no internal activity in the Mind.

—SIT comfortably and just look for a while, doing your best to just be with the *object* and commit it fully to your *Awareness* without analyzing any reactions or associations.

—CLOSE your eyes and IMAGINE a copy or facsimile of the *object* firmly in your Mind's Eye.[*]

—IMAGINE the facsimile *object* as closely to the original *object* as possible.

---

[*] When realized at *Alpha Awareness* levels, these types of processes (indicating the "Mind" or "Mind's-Eye") may be conducted from "Spirit Vision" as an advanced application.

—WILL the brightness of the *object* to increase; then WILL the brightness to decrease; then increase again.

—WILL the color of the *object* to change; make it blue; then turn it red; then change it to green; then return it to its original color.

—OPEN your eyes and FOCUS your *attention* back solely on the original *object*.

—WILL the brightness of the *object* to increase; then WILL the brightness to decrease; then increase again.

—WILL the color of the *object* to change; make it blue; then turn it red; then change it to green; then return it to its original color.

—CLOSE your eyes and make facsimile copies of anything that enters your Mind—each time an image or thought enters your mind, IDENTIFY the *image*, then copy it and duplicate it next to the original. DUPLICATE the image again. DUPLICATE the image again. Keep doing this several times until all *Awareness* of the image dissolves.

—OPEN your eyes and FOCUS your *attention* back solely on the original *object*.

—IDENTIFY the *object*, then make a mental facsimile-copy of it in your Mind and duplicate it a few feet in front of you.

—DUPLICATE the image again, placing it behind you.

—DUPLICATE the image again, placing it on your right side.

—DUPLICATE the image again, placing it on your left side.

—DUPLICATE the image again, placing it below your feet.

—DUPLICATE the image again, placing it above your head.

—DISSOLVE the imagery and IMAGINE your physical body is enshrouded in a *sphere of light* before ending the session.

UNIT SIX

SELF DETERMINATION

THE ROUTE TO
SELF-DETERMINISM & BEYOND

Education and practical methods provided in this volume lead a *Seeker* through present discoveries of ongoing research efforts by the *NexGen Systemology Society (NSS)* toward *Self-Actualization* and spiritual evolution of the Human Condition. We are working consistently toward a new state of realization for a futurist form of humanity called *Homo Novus*—the *NexGen Human*—that has solved the labyrinth of worldly systems inhibiting *Self-Determinism* and traversed the spheres of existence leading to greater and greater realizations actualized by a *Seeker* increasing their personal vibrations closer and closer to the highest *Alpha* states we can theorize from *beta-existence*.

> *Self-Determinism* is <u>personal control</u> of WILL
> to direct *Intention*.*

Defragmented *Self-Determinism* is a necessary prerequisite for personal demonstration and development of all other "higher" capabilities of the *Alpha-Spirit*—for example: to postulate the *Games* and *Logics* of the Spiritual Universe (AN); to design and develop other beta-*Universes*; to imagine and solidify organic systems and *bodies* to experience those universes... The lore remaining today as passed down from the Ancient Mystery School is quite generous in its esteem of those qualities possible at highest realizations of the Human Condition—far and beyond the *beta-programming* instilled by "cause and sequence" or "default" experience that we identify with *beta-existence* in *this* lifetime.

This present cycle of *Systemology* work (within this volume) provides tools for a systematic *reduction* of *imprints*, an increase of *certainty* toward personal management, and critical

---

* Paraphrasing a line from one the present author's first underground publications from 1998, *"The Sorcerer's Handbook,"* originally released to the underground under the pseudonym of "Merlyn Stone."

developmental steps that appear universal to all effective methods lending to *Self-Actualization* and spiritual evolution. An additional run-through of this cycle of work will accomplish additional *reductions* of the barriers and increases of the *freedoms* available to the *Seeker* as they operate in *beta-existence* for this present *Game of Life*. We have demonstrated a progressive sequence of incremental "gains" for the *Seeker* on the *Pathway to Self-Honesty*—and no doubt there have been many "peak experiences" of *enlightenment*, *Gnosis*[140] or *apotheosis*[141] along the way—each providing (necessary) personal confirmations that inspire and incite further efforts on the *Pathway*. These individualized "gains" and significant *"ah-ha"* moments are only determinable for-and-by the *Seeker* themselves. It should be noted: there is no point during the *Processing* in this book where a *Seeker* will suddenly sprout wings and fly away—but it is maybe something to consider when creating your next body for another planet or universe.

We have previously explored many varieties of *emotional encoding*, *imprinting* reinforcement and *programming*—all of which are demonstrated to restrict *Self-direction* and dampen *Awareness*. In order for the *Seeker* to achieve even greater gains from this point, and extend reaches to further points on the *Pathway*, it is necessary to dig into even deeper crevices and wide-sweeping blankets that have yet to be pulled into the clear light.

Effective *Self-control* and personal management of environment is a quality of *Self-Determinism*. The ability to effectively decide upon (intend) and execute an action (effort) is a quality of *Self-Determinism* when not inhibited by *fragmentation*, other *programming* or an *enforced reality*. When an individual

---

140 **gnosis** : a *Greek* word, meaning *"knowledge,"* but specifically "true knowledge"; the highest echelon of "true knowledge" accessible (or attained) only by mystical or spiritual faculties.

141 **apotheosis** : from the *Greek* word, meaning *"to deify"*; the highest point or apex (for example, of "true knowledge" and "true experience"); an ultimate development of; a glorified or "deified" *ideal*, such as is a quality of *godhood*.

is able to freely *Self-direct* all *intentions*, *thoughts*, *emotions*, *efforts* and *actions*, they are maintaining a state of *Self-determinism* in *Self-Honesty*. Because *Self*, when considered from the perspective of the *Alpha-Spirit* is a "constant" presence of ZU, the only inhibition to a total actualization of this state during this lifetime is *fragmentation*. We must look beneath the surface of *"moves"* and *"counter-moves"* in this *Game*. Beneath the *efforts* of others (which are *counter-efforts* to us), restraining and invalidating *thoughts* and *beliefs* of others (which are *counter-thoughts* to us), and the *enforcements* and assaulting *actions* of others (which are *counter-actions* to us), we are left primarily with *agreements* and *decisions* that we personally have decided *to be* a Reality. These include ways in which we actively contribute to our own *fragmentation*.

SYMPATHY-IMPRINTS AND EXCUSES

Very few *imprints* so successfully eclipse[142] a *Self-determined* WILL (5.0) as quickly and effectively as the *sympathy-imprint* and *self-denying excuses*. These types of imprints are successfully reinforced into personal reality by agreement—and usually during times of perceived "stress" or confusion.

> The first type—*sympathy-imprints*—emerge when *others* falter in their certainty of *self-determinism* <u>and</u> we *choose* to *Identify* with them.

> The second type—*self-denying excuses*—are reinforcements of our faltering certainty of *self-determinism* when we *choose* to *Identify* with former *sympathy-imprinting* and thereby *"deny"* ourselves present certainty.

Sympathy-Imprints are *emotionally encoded* into the *beta-personality* or "character" of an individual. They are used in place of *actualized* "knowledge" and "certainty" during times when an individual is uncertain of their ability to *Self-direct* a definit-

---

142 **eclipse** : to cast a shadow or darken; to block out or obscure a comparison.

ive decision—such as "yes" or "no." They will rely on the presence (or artificial manifestation) of a *sympathy-imprint* in order to provide displacement of a *Self-determined decision*, and hence: the *"excuse."* These are low-level energy attempts to continue one's own "existence"—yet affirming and agreeing to shortcomings of such existence, hence: *"self-denying."*

> A *Sympathy-Imprint* is used to *excuse* personal failure.
> The *excuse* reinforces *emotional encoding* of failures.

One reason that *Sympathy-Imprints* are so strong—and sometimes difficult to *defragment*—is because an individual is usually in *agreement* with the Reality *facets*, such as a familiar environment, a fond person, &tc., and is primarily "conscious" of activities when *imprinting* and later *reinforcement* are taking place. Although "conscious agreements" are made —they are made in *fragmentation*. They are made in the absence of higher frequency operation of WILL and are instead treated from a low-ZU emotional level.

The very idea that an individual "must drop down in vibration in order to assist" a person that is suffering needs to be *"processed-out"*—especially for those who prize themselves on their "outward demonstrations" of exaggerated suffering, which is, itself, a use of *sympathy-imprinting* and reinforced *self-denial*. These low-levels of experience spread like a disease. The Western world is programmed to so heavily "identify" with their fellow "Humans"—foregoing any individualism or *Self-direction* to actually and effectively assist humanity—and instead wallow along with the rest of the herd self-validating their powerless position as victims. This is not the way of *Self*, *Wizard*,[∞] or even the *Master*[‡]—it is a definitive

---

[∞] *Wizard*—NexGen Systemology "Wizard" Levels I–III are *Grades IV, V* and *VI* as developed by the *Mardukite Research Organization* in conjunction with the *NexGen Systemology Society (NSS)* for the *International School of Systemology (ISS)* "Flight School / Pilot Training" issued by the underground *"Systemology Air Command (Elite Force)"* developing in 2020.

[‡] *Master*—Mardukite Systemology "Master" Course materials cover the

sign-post for a *Seeker* still not yet fully *beta-actualized*.

As a method of soliciting validation from others, *sympathy-imprints* are automatically reinforced whenever found to yield "positive" results. By positive, we do not mean morally "good" versus "bad"—rather it means that there is an "additive" *effect* to the environment. A "positive suggestion" (in terms of "hypnosis") is not a "happy command"—it is the intention for something *to be*. An individual discovers *sympathy-imprinting* after failing to follow an initial command from *Self*—thereby failing to "positively" execute control or management of Reality by traditional tendencies or reasoning. At this point—relative to the perceived/experienced emotional intensity of the failure—an "excuse" or "self-denial" is employed to enact *a "positive" effect* that moves an *Identity* forward in "getting by just one more moment" of validated existence, and this is when *sympathy-imprinting* is reinforced as behavioral conditioning (based on its observed "effectiveness"). *In actuality—from the perspective of pure logic—* this "appeal to pity"[*] is a very low-energy method of persuasion used only to accumulate Reality agreements from others based on their own tendencies/ programming for "sympathetic compassion."

Sympathy[143] is aberrative. It is falsely assumed to be a requirement for "understanding." The two are *not* the same. The demand for "sympathetic compassion" from others—even when only "passive-aggressively" expressed in our vicinity—is a favored "control mechanism" used by those operating at very low-energy frequencies on the *ZU-line*, such as *(0.6) or lower*, in a crude and primitive attempt to sustain their survival via assistance or cooperation from others. This can actually begin at any point on the scale from the

---

    entire scope of *Grades I, II* and *III*.
* "Appeal to Pity"—an informal fallacy; *argumentum ad miscericordiam*.
143 **sympathy** : a sensation, feeling or emotion—of anger, fear, sorrow and/or pity—that is a *personal reaction* to the misfortune and failure of another being.

"pain threshold" (approximately "1.8") on *down*, Such efforts are meant to "evoke" or "conjure" a certain *emotional effect* in others—thereby allowing a "suffering individual" to resume some semblance of *cause* in *beta-existence*.

Any manner of "sympathy" offered serves to further *fragment* both (or all) individuals directly involved in the energetic interaction. Some individuals find themselves in cycles of this behavior for no other reason than that the *conditioning* has been positively reinforced by others providing the sympathy attention-energy. This can become an energetic life-support drug-like dependency on external energies to simply "ride the waves out" in *Life* at the bottom end of the ZU-line. When such levels of *Awareness*—or lack thereof—are maintained, any attempts at actual effective productive assistance or "help" will undoubtedly be rejected, which only serves to provide additional *fragmentation* to the individual providing *efforts* to be of *assistance*. This entire matter is a very boring game of juggling grenades—which is why "prevention" (a byproduct of increased *Awareness*) is emphasized in *Systemology* above all else.

When an individual fails at some effort in their own experience, whether toward a creative goal or something more tangible as sustaining their existence, there is an instant reduction in *Awareness* on the ZU-line. Of course, for the well-adjusted *Self-Actualized* individual, this dip is momentary and incidental—and they are able to quite quickly snap back to their normative mode. If, however, the *Awareness* level drops and remains fixed and *emotionally entangled* below (2.0), *sympathy-imprinting*—or stimulation and reinforcement of existing *imprinting*—and its corresponding "reactive-response mechanisms" are engaged by the RCC. This is an attempt to solicit assistance and communication (energy) from others by invalidating their state/condition and pulling them down to symbiotic[144] vibrations. This "primitive social mechanism"

---

144 **symbiotic** : pertaining to the closeness, proximity and affinity between two beings that are in mutual communication or maintaining

may have originally, at some distant time, served some degree of actual survival value among populations sharing equally low-level *Awareness* and poor language skills—but it certainly no longer serves the newly evolving Human Condition, especially on the *Pathway to Self-Honesty.*

With the absence of higher-level thought (beta-Awareness) and/or an actualized WILL to express the *Beingness* of *Self* (*Alpha-Spirit*), an individual turns over the "source" of their failures to some type of "Self-defeatism." In the face of a failure, when others are not (or cannot be) misappropriated as the cause of our failures, the RCC will provide all the causal justification necessary to generate an illusion that the original calculated *efforts* were not flawed, because something else "made" such and such happen. And although not directly an emphasis of *NexGen Systemology*, certain observable physiological health issues—which some have referred to as "psychosomatic"—actually do manifest from *agreements* of *Self-denial* and "fault" that were not *Self-Honest* in their intention. In short: we can *realize* ourselves "sick" and then *actualize* it. And this is one instance when higher grades of *Systemology* education already demonstrate to a *Seeker* very clearly that the "universe" as some put it—or else the "Cosmic Law" governing ZU activity—responds very loudly and surely to *Self-Honesty*, and *any* fragmentation thereof.

MANAGING SYMPATHY & LOW-ENERGY

Assisting others in elevating their *Awareness* and promoting greater success for all *Life* and *Existence* to the furthest manageable reach of influence is a very high-energy *Self-purpose* in *beta-existence*. The *Effort*, of course, may only be effectively and safely applied from those that are, themselves, maintaining high-energy vibrations of personal ZU. This is stressed consistently because any personal weaknesses and *fragmentation* is likely to be used and/or stimulated during low-level

---

mutually validating interactions.

interactions. And the modern world seems to maintain a certain amount of unhealthy control simply by keeping the population frozen in its *imprints* and *emotional encoding*—constantly evoking and inciting this low-level direction of our *Awareness* and focused attentions.

Sympathy—and its related circuits/cycles of activity—is demonstrated only at lower levels of ZU-energy; moments when there is an absence of *actual* and *effective* assistance and "help." While there is absolutely nothing wrong with assisting others, there *is* something wrong with succumbing to the same states we were "hoping" to assist. High-energy individuals are able to assist and direct energy toward others without falling prey to such circuits of energy—and without further propagating these circuits of *fragmented* energy in their own lives (and the lives of others). This is a challenge for many individuals experiencing the Human Condition today.

> Inherent programming and social conditioning within the "standard issue" state of the Human Condition actually uses philanthropic[145] tendencies of a "decent" individual as "facet fuel" to further *fragment* that individual in the low-level systems for their *efforts.*

This is one reason why many on the *Pathway to Self-Honesty* find it difficult to *actualize* beyond this point into even higher "grades" of development: because it runs so contrary to the basic "just do whatever you can to get by" mentality that is inherent within "standard issue" programming.

Personal *fragmentation* is not simply attached to our "failures" but to how we handle the truth and responsibility of our failures—and this includes our evaluation of failures of others. There is nothing wrong with making calculated attempts toward an *effect* and "failing" to reach the intended results so long as the entire process is *Self-directed* and free of existing

---

145 **philanthropy** : charitable; the intention (or programmed desire) to generously provide personal wealth and service to the well-being and continued existence of others.

fragmentation—and *Self* may actually glean "true information" about its *efforts*. However, when an individual is succumbing to very low-energy states as a result of perceived failure, the RCC systems engage to fill in gaps of knowledge that prevent *Self* from understanding and realizing its own way *out*. If the individual determines that they do not possess the ability to assist themselves, they will demonstrate (and reenact or dramatize) facets of their *sympathy-imprinting* as a reactive-response survival mechanism—to "appeal to pity" from the environment. This is the: "Oh, don't hurt me; I'm already down." Or, the enemy that faces its final demise and switches on emotion as a tactic to regain the upper hand. We've all seen demonstrations of this in our lifetime.

A *Self-Actualized* individual may still make mistakes and experience "failures" in *beta-existence*—we are not suggesting otherwise. Mistakes and failures are simply results of miscalculated *effort*. This may occur when the "thoughts" going into these efforts are *fragmented*—or when the information received (input) about our environment is *fragmented*. In other words:

> the *Alpha-Spirit* is capable of making the best decisions and resolving consequences when *actualized* beyond a point of *beta-fragmentation*.

However, up unto this point, a *Seeker* on the *Pathway to Self-Honesty* must "practice" *realizations* of this *actualization* until it is "automatic." Automatic tendencies will always be found in the systems—there is no concern about this so long as systems are *defragmented* and *Self* has directed their autonomy.

As an individual becomes accustomed to operation of *sympathy-imprints* for their own purposes, the level of *Self-determinism* that is *realized* diminishes. This only can happen when an individual makes a conscious decision to relinquish that control and misplaces *cause*. Therefore, a thought is directed to solicit the sympathy of others in order to ride the lowest existence that will still carry an *Awareness* into the: "if I can get through just one more day" type of reasoning. Of course, nothing along this mode of thought and action can effectively contribute to actualized longevity unless a person is basically put on permanent "energetic life-support" by their environment. This is where the "coddled individual" becomes *helpless* in resolving and managing their own personal affairs. They may *appear* to get along just fine—so long as they are fully dependent on management by others—but get them alone and in touch with their own affairs as *Self* and the truth will be revealed. Certain situations generate this in children (or anyone) under the authority of parents and other "trusted" adults—and elements may also be found in *jails* and *military* establishments; anywhere authority is enforced.

—IMAGINE your physical body is enshrouded in a *sphere of light*.

—REVIEW moments during your lifetime when you affirmed *illness* or *personal inability* as an *excuse* to avoid going somewhere you didn't want to *go*.

—ANALYZE your intentions motivating each of these instances. *Where didn't you want to go? Why didn't you want to go? What facets are encoded in this imprint?*

—REVIEW moments during your lifetime when you affirmed *illness* or *personal inability* as an *excuse* to avoid completing a task you didn't want to *do*.

—ANALYZE your intentions motivating each of these instances. *What didn't you want to do? Why didn't you want to do it? What facets are encoded in this imprint?*

—REVIEW moments during your lifetime when you affirmed *illness* or *personal inability* as an *excuse* to avoid participating in a role you didn't want to *be*.

—ANALYZE your intentions motivating each of these instances. *What didn't you want to be? Why didn't you want to be it? What facets are encoded in this imprint?*

—REVIEW moments during your lifetime when you affirmed *illness* or *personal inability* as an *excuse* to avoid accepting something you didn't want to *have*.

—ANALYZE your intentions motivating each of these instances. *What didn't you want to have? Why didn't you want to have it? What facets are encoded in this imprint?*

—REVIEW moments during your lifetime when you exhibited *illness* or *personal inability* as an *effort* to gain *sympathy* or *assistance* from others.

—ANALYZE your intentions motivating each of these instances. *Who did you solicit sympathy from? What was the motivating failure? What efforts did that person provide—and what did they do?*

—RESURFACE any *sensations* or *facets* associated with each of these event-instances. ANALYZE any emotion expressed by others—*and* your own emotional response to their *efforts*. ANALYZE any statements verbalized by others—*and* your own thoughts associated with their *intentions*.

—REVIEW moments during your lifetime when you expressed *sympathy* or *sympathetic effort* in a response to *assist* another being's *Beingness*. [This includes any living organism, physical object or creation.]

—ANALYZE your intentions motivating each of these instances. *What failures motivated a response? Who/What did you provide sympathy to? What efforts did you express—and what did they result as?*

—RESURFACE any *sensations* or *facets* associated with each of these event-instances. ANALYZE any emotion you expressed in your effort—*and* the other's emotional response. ANALYZE any thoughts associated with your *intentions—and* any verbal statements expressed by others in response to your efforts.

—IMAGINE the scene of an *injured child.* —REVIEW your level of *sympathy* and sympathetic effort or emotional response. —ANALYZE how you could *help* this situation.

—IMAGINE the scene of an *injured parent.* —REVIEW your level of *sympathy* and sympathetic effort or emotional response toward. —ANALYZE what you could do to *help* this situation.

—IMAGINE the scene of an *injured beloved pet.* —REVIEW your level of *sympathy* and sympathetic effort or emotional response toward. —ANALYZE what you could do to *help* this situation.

—IMAGINE the scene of a *burning forest.* —REVIEW your level of *sympathy* and sympathetic effort or emotional response toward. —ANALYZE what you could do to *help* this situation.

—IMAGINE the scene of an *injured stranger.* —REVIEW your level of *sympathy* and sympathetic effort or emotional response toward. —ANALYZE what you could do to *help* this situation.

—RECALL the last time you successfully assisted another being and felt the relief and satisfaction of *directing* the *cause* of *effects*.

—IMAGINE your physical body is enshrouded in a sphere of light. End the Session.

WILLINGNESS AND SELF-DETERMINATION

Systemology Processing demonstrated with the present volume is *light*, but effective. Because of the objective "one-sided" demonstration as a static book, we have skirted across considerable territory concerning artificial *fragmentation* and accumulation of *character personality* that becomes identified with *Self*. As the mechanistic *programming* and *imprints* are reduced one by one, the *Seeker* takes one step closer to the full certainty, ability and *Self-determinism* to direct personal management *efforts* surely and securely. *So what* if we make mistakes? We take responsibility and own it—learn something about that specific event—and move on with future application of *efforts* without a concern that we have failed at something before and may do so again. But, don't even let a single particle of that *antique dust* of yesterday remain on you to slow down your motions toward tomorrow. It won't impress anyone—you wont get any additional commendations for superfluous struggle. If your *Awareness* dwells in the hurts, smites and failures of yesterday, then a *Self-Determined* future will prove impossible. *Get up; dust off; keep going!*

When we consider all the "esteem" and "pomp" that is added to this idea of "accumulating experience as we *get older*," we should expect some fairly incredible things to happen. At every turn of life in this society, we are assured that authority knows best, our elders are always right and that we will just *know better* when we get older. Of course, in *Self-Honesty*, we know that this is *bull*. Our previous generations have not been observed to know best; our elders succumb to apathetic stagnation, because they realize that some immanent savior or imminent state of enlightenment never arrived; and in the end, when enough experience has been accumulated and the *organism* can no longer function as *cause*, the last steps are to hold on dearly to whatever *personality programming* has carried them at least that far, and it winds up serving as the very anchor to pull them the last of the way down.

Now, of course, we have all seen the biology films and watched the cycles of Nature take their course. Organisms grow, live, die and are reformed into energy to fashion new forms of organic life. This happens continuously and is even a part of the very cellular and genetic makeup of the *vehicle* you are using as an "avatar" for *beta-existence*. It has its own memory, inherent in the slow progression of its own cellular evolution and development—and YOU as the *Alpha-Spirit* and "I" that is the actual true spiritual YOU, also have a memory and a true personality that you have chosen to develop as an individuated spiritual being, which has experienced many lifetimes as well.

The entire issue—and the solution *defragmentation* provides—is returning *Awareness* to be as *one* with the actual spiritual *Self* to determine personal management of *Self* and experience of this *beta-existence*. We are often led to assume this is "automatic" but it *is not!*—and what's more: it appears to qualify a primary difference between the actualized figures of legendary and historic renown versus what we see demonstrated as "passable" for the *standard issue Human Condition* in operation today. This is one of the key areas that *Systemology* works to resolve for the next evolution. Certainly anyone with even *some* degree of actualized *Awareness* would agree that this is the case regardless of any other semantics—philosophical, mystical, cultural or religious background—is brought to the cycle suggested in this volume and any other related *Systemology* work.

At this juncture of the *Pathway*, our instruction and processing turns toward the subject of *Willingness* and *Considerations*. In previous lesson-chapters and demonstrations of *Self-Processing*, we have danced between the high-tides of the matters that are most "present" to us and those which are "distanced." Primarily we have been concerned with things that have happened *to* us—and the type of *imprinting*, *facets*, *associations* and overt *programming* that we have experienced in this lifetime. It is interesting, in retrospect, that the accumulation of all these *solids* and inhibitions of *Awareness* should be so highly prized among the ignorant (when you consider the way people go on about their *sufferings* and *experiences* and "*character-building*"), because as you have seen with even the progress you have made in the short time already spent on the *Pathway*, that peeling these layers away and reclaiming your own personal power is *really tops*.

A *Self-Determined* individual is completely certain in what they are *willing* to do—after systematically analyzing (mentally exploring and evaluating) <u>all</u> *considerations*. It is only when the *imagery* associated with these *considerations* is automatic, *fragmented* (influenced by *imprints* and *programming*) and/or assumes an emotional "life of its own" that the *Seeker* has issues. So long as the *illusions* and *thoughtforms* are completely under *Self-control* to create and dispose of for *consideration*, then there is no concern of *fragmentation*. In fact, it is the entire purpose of the ACC (7.0) and "Alpha Thought" ("Spiritual Mind-Systems") to create and dispose of reality considerations at *will*, in order to evaluate the nature of the very *Will-Intention* (5.0) that *is* put forth, projected or expressed further down the ZU-line into more "solid" manifestation.

When *Awareness* is operated at high-frequencies—during *processing sessions* and every-day living—it is much easier to make large sweeping *scans* of possible considerations and all aspects related to any events—which is why we use the very term *Awareness*. By increasing actualized *Awareness*, a *Seeker* is increasing what is "possible" to be known as Reality—the nature of *realizations* possible. And as we know, everything *actualized* as an *Awareness* must be first *realized*. An individual has full range of *consideration* to the extent of *Self-Honesty*—and that range of *realization* contributes directly to what an individual is <u>willing</u> to do as *Self-determined Will-Intention*. As you can see, the semantics of basic language have played with us again,* since the condition, qualities and properties of WILL (5.0) is literally being referred to here as *Willingness*.

CONSIDERATION AND WILLINGNESS

A *Self-Honest* individual should have no difficulties or reservations exploring an entire *infinity* of potential *considerations* for any and all given situations, experiences or interactions that they are faced with—or could ever be faced with. They also should have the ability to recall any *Awareness* and attention-energy back from any of these *self-creations* and *thought-forms* without attributing to them any actual emotional energy or effort. In other words:

> A *Self-Honest* individual should be able to consider *anything* without *fragmented responses* or *(re)-actions*.

A person *should* be able to create and consider *any* image or

---

* See *Tablets of Destiny* concerning hidden meanings in language.

course of action (sequential images) in the Mind-Systems without becoming an *effect* to it—without it becoming the *cause* of an automatic response-reaction. That is why *Self-Control* is stressed so highly: because once upon a time, a more actualized humanoid form did present itself on this planet that *could* WILL effects much more swiftly and effortlessly than we practice today. These beings dampened *Awareness* of such faculties in the Human Condition because without *Self-Honesty*, they are simply too "destructive." Hence, now, in order to actualize them from an experience of the Human Condition—moving up the "Ladder of Lights" or *Pathway*—there are a lot of "checks and balances" of that progress, making certain our programming and ability is cumulatively maintained before an equal responsibility for power is blatantly demonstrated. These systems were originally put in place to the fullest extent of "Cosmic Law"—with every facet exploited to its limits.

Another aspect that can actually cause turbulence for those without higher levels of *Self-Determination* are the "obsessions" and "harbored feelings" that *solidify* as a direct obstruction to *Self-Honest* experience of *beta-existence*.

> "Obsessions" and "fixations"
> solidify *Awareness* on an
> "unrealized consideration."

This does not happen when an individual mentally exhausts all considerations with analysis and evaluation—but *Awareness* will always get "hung up" on a *maybe*. This isn't typically a problem when we apply our highest faculties to determine the highest route to chart our course of *efforts* and *actions* on with *certainty*.

> *Consideration* is everything that you *could realize possible*.
> *Willingness* is everything that you *would actually do*.

> A *Self-Honest* individual should be *able*
> and *Willing* to fully *Consider* anything.

The *Willingness* and *Consideration* you *can see* expressed speaks volumes of what an individual covertly conceals.

Willingness is not the same as "all that is possible"—it is a *Self-directed* collection of applicable *efforts* that an individual is *willing* to *intend*, but only *after* exploring all *considerations* of what is possible. Therefore, we are touching an even deeper level of "personality" and "character" than is otherwise explored in the present cycle of materials. A *Seeker* is here brought to scan, review and examine the widest range of *considerations* possible—and if this is not a very wide range, we suggest a consistent reapplication of the *processing* cycle demonstrated in this present volume. Here at this juncture of the *Pathway*, the *Seeker* will undoubtedly begin butting up against whatever "self-imposed limitations" have been placed on *Alpha* qualities, such as: creativity, imagination, consideration and willingness—all of which manifest prosperously in *beta-existence*, when they are able *to be* expressed via the *clearest channels* of ZU-energy.

When an individual is highly actualized, the range of consideration is wide and the inhibition for expression is slim. That does not mean that a person just *does* whatever they want without a regard for consequence—on the contrary, an actualized individual will *Self-determine* only the best courses of action in any given situation. This is only possible when there is no *emotional* or *programmed* fragmentation that inhibits all considerations for intention. An individual's demonstration of WILL is only limited to the extent of their considerations—and then the ability to evaluate and analyze those considerations as courses of action. This resolves the double-edge sword inherent in "getting what you asked for"—because whatever you do *decide* to place your attention-energy, WILL and intentions upon, you can be assured that it will become *real* and *solid*—at least *to you*. Even when a personal agreement to some reality is shared by no one else, you can be assured it is *to be* Reality for you.

By removing erroneous programming or encoding, the

Alpha-Spirit may retain *Awareness* of "total consideration" and apply the greatest certainty to all *Self-determined* expressions of WILL.

This is the only sure way to plot (or *"pilot"*) the course and navigate the way ahead—*to be* the *Self-determined* cause of the best course of action that leads toward a continuous *Self-directed* spiritual existence that is independent of the Physical Universe and still maintained in full *Awareness* by *Self* during the experience of this incarnation in *beta-existence*.

When we use the *processing command line:* CONSIDER, the *Seeker* is to repeat that "line" of *thought* until all possible *considerations* have been exhausted. This also includes "imaginary" considerations that surface to mind. Even the concept/word of "consideration" may need to be *processed* before such an exercise can be actualized. The reason being that modern society has *enforced* its own *programming* concerning (and limiting) the mental faculties of *consideration*. For example, when we are instructed by authorities to "consider others" they mean for us to be controlled in our exploration of free thought and wide consideration by placing "mental taboos" on anything that is *outside* the domain of standard issue normalcy. Of course, the moral implication regarding contemporary "consideration" in the modern world is to always *extrovert Awareness* toward others, social standards and duplication of what is already established in agreement. We are not implying that a person should not demonstrate the highest regard to all *Life* in existence—what *is* implied, concerns the *imprinting* and *emotional encoding* that retains a hold on abilities to actualize higher faculties of *Awareness* (ZU) than are encouraged or easily replicable in *beta-existence*.

THE REUNION OF AWARENESS AND SELF

The journey on the *Pathway to Self-Honesty* reveals deeply laden ancient secrets about the Human Condition and the systematization of *beta-existence*. It unlocks and unfolds latent faculties that are embedded behind the "emotional states" and "mental consciousness" that we ascribe to the extent and flaw of the Human Condition—and yet, we have seen in many turns of our own life and all throughout history that there *is* something *more*, and we do not necessarily have to "die" to prove it. We <u>can</u> *realize* this "heaven" on "earth" and *actualize* it during *this* lifetime.

Fragmentation is the separation of "consciousness" from *Self*.

Mental imagery and *beta-thought* is primarily bound to "reactive-response" *conditioning* and other *imprints* that constitute a "beta-personality" and "artificial character." This is assimilated as a *facsimile* of an archetype—the true *Identity* (or *Alpha Personality*) of the *Alpha-Spirit* or *Self* in its true spiritual state. This means that apart from *Self-Honesty* and free range of *Alpha Thought* tapped *outside* or *exterior* to the Physical Universe (KI), <u>all</u> *beta-thought* and "inclined considerations" will be restricted to *preexisting* agreements to Reality. Once *Awareness* is effectively reunited with a true *realization* of the *Alpha-Spirit*, a *Seeker* is able to access a wider range of "consciousness activity" for their *consideration*. This is a pivotal milestone in reaching the ultimate goals of *Systemology*.

While developing states of increasing *Self-Honest Awareness*, the *Seeker* then *realizes* their *Awareness* (or "seat of consciousness") from exceedingly higher points than found in existing contemporary textbooks on science and history and other *exoteric* academia. In *NexGen Systemology*, this meta-human state of *Awareness* marks an inception of a new stage of "spiritual evolution" for the Human Condition, which we have called *Homo Novus* (or *Homo Novis*) for public purposes—as we are certain that such *NexGen* individuals will indisputably demonstrate that they possess a much greater level of *Actualized Awareness* than their fellow members of society. Internally, we are referring to the development of these states as "A.T." for "*Actualized Technician*."[∞]

[∞] "AT"/A.T. *or* "Wizard Levels" of *NexGen Systemology*: *Grades IV, V*

We are already—and have for some time—been witness to a slow development of this potential genetically inherent in the *Gen-Y-Millennials* (born 1982-2000) and this should be equally cultivated for the up-and-coming *Generation-Z* (born 2001-2019) that are entering into a world that is greatly uncertain of its own self and its own past...then alone where it is going. Who knows if we will even see a full term of another generation, or if *Homo sapiens* have hit "generation black" and will be altogether replaced by *Homo Novus*. By 2038, we are likely to live in a world dominated by our ATs, AIs, or with no healthy Earth planet to live on at all.

> The *Pathway* to a successful tomorrow is not paved in bricks forged from yesterday's failures. It is up to those with greater vision and higher views to *pilot* the course of humanity's future in *Self-Honesty*.

The present volume of material should have, if nothing else, demonstrated very clearly that the extent of our *beta-existence*—when operated outside of *Self-Honesty*—is determined by the contents of our databanks of *beta-Awareness*, which is to say "*beta-banks.*" The contents, which we have *resurfaced* and *recalled* into the light considerably, are found not to be the true possessions of *Self*, but of those which have been *enforced, conditioned, encoded* or *programmed* by our *beta-existence*. The greater the contents of our "*beta-banks*" the greater the *fragmentation* inhibiting clear view and expression of *Self*. The *facets* used to hold the "*banks*" together coherently are an amalgam of *fragmented experiences* and the *suggestions from others*, but with little (if no) content deposited *intentionally* by *Self*. This is because, unless otherwise *fragmented*, the *Alpha-Spirit* does not intend on becoming an *effect* of its own *cause*—it prefers a fluid and etheric state that is easily transmutable. This is not *realized* during this lifetime while *Awareness* remains solidified and fixed in *effects*.

---

and *VI*. The term relates back to early years of development and the "*Reality Engineering*" lecture-series in combination with the present semantic regarding *Systemology* "techniques" and "technology" as "*Tech.*"

Prior to the recent integration of *Mardukite Systemology*, the modern traditions, religions and "New Age" methods—as they are currently *realized* in the 21st Century—have all only demonstrated moderate degrees of effectiveness, and generally only with a few actualized practitioners that have realized any true *Self-Honest* benefit from their practices of focused concentration, meditation and ritualized magic or mysticism. We have obviously been in need for something more and it must come down from the *Alpha-Spirit* to be effective or we will be forced to remain enslaved to the memory of this lifetime and former lifetimes of the genetic body that undeniably carry the *imprinting* of failure accumulated from countless lifetimes even before this one. At *Grade III*, this cycle of *processing* is likely to perhaps already have triggered a glimpse at some former cellular genetic programming of the body—or even another lifetime of the *Alpha-Spirit* that may not even be restricted to this planet! Such matters may be taken up in later Systemology volumes.

WILLING "TO BE" A SELF-DETERMINED SPIRIT

Systemology Processing has been described by some as: the "discovering"—or, more correctly, the "uncovering"—of the most basic personality and actualized *Awareness* of the *Alpha-Spirit*. This is our true *Identity* as *Self* or "the Spirit that is *willing*" the body to think, feel and act, when it is *Self-Determined* in these responsibilities and faculties. The systematic methods and education demonstrated within the present volume are all the result of personal exploration by the present author—and the research organizations connected to the same—which have been found effective to directly lead toward "recovery" of the spiritual potential latent in the Human Condition.

The *Pathway* is inherently "graded" to the extent that an individual is capable of developing incrementally toward *Self-*

Honesty and even higher states of spiritual *Self-Actualization*. Any barriers and challenges to success are self-imposed—meaning that the quantity or quality of *fragmentation* is very unique to an individual, their current and previous lifetime from a genetic/physical perspective. This means that to be effective, *Self-processing* provided in this volume is meant to put *Systemology* on the radar of public interest in lieu of any previously established organization base at the time of writing/publishing this "self-help" volume.

Processing presented within *Crystal Clear* is quite valid and effective when employed alone—however, during the time spent testing and experimenting with this work, it became very quickly apparent that the swiftest and surest progress was attained by *coordinated efforts* that involved professional/expert *Piloting* of a *Seeker* through various terrain on the *Pathway*. This is not to slight out any individual *efforts* applied, which are certain to have already found validation by this point of the cycle, but it should also have become clear that there are certain limitations to delivering this material as an introductory-level book confined to satisfy the specific conditions assigned to the present author: "firstly, a self-help workbook manual of *Systemology* that may be applied either alone and independent of assistance; and secondarily, a basic outline to introduce *piloted* assistance to the *Seeker* so long as it does not conflict with the first condition."

This cycle of work develops certainty toward personal management and responsibility. It should be evident by now that we cannot simply run around the world agreeing with the Reality of every body we encounter—nor can we simply run around arbitrarily enforcing agreements on others. Our *Systemology* should demonstrate, above all else, that there *is* a *systematic* balancing of forces constantly in motion, and the best course of action is to walk the middle *crystalline* path and avoid adhering too closely to the extremes. Adhering to one extreme in *processing*, the *Seeker* would find that they have "blown the Mind wide open" and would immediately fall out of contact with *beta-existence* and other matters of the Physical Universe (KI). At the other extreme end of this spectrum, we find *no processing* and the more commonly held tendencies held by the standard issue Human Condition to consistently add more and more layers of confusion to perceptions held about the *Identity* of *Self*, and eventually succumb to the lowest state of being, which is to be "total *effect*" of environment and others. Therefore the spirituality behind *Systemology Tech* is best equated to what is esoterically known as the "*Crystal Path*" and in other philosophies as the "*Excluded Middle*." This is not directly present in any other futurist 21st Century methodology.

As a *Seeker* cycles through the *processing*, the turbulence of *Life* created by *fragmentation* is lessened. The emotional *imprinting* and limitation fixed to our beliefs is *systematically* released bit by bit. In doing so, the latent potential of the Human Condition is "unfolded." It is important also that a *Seeker* is only undertaking the *Pathway* and performing the *processing* at a pace that is not, in itself, a source of turbulence. It is perfectly acceptable to push one's limits to greater frequencies than are typically encouraged in a sluggish world, but it is important to remain within the "tolerance band." There is a glimmer of *Self* that is always accessible and in existence at (7.0) regardless of wherever else active *Awareness* may be fixed. Certain creative processes suggested are meant to allow a *Seeker* to catch that *realization* of higher *Awareness* as a much higher and more effective point to fix the *attentions* to, meanwhile operating the management of *beta-existence*. These two states are only separated in "consciousness" and are otherwise one continuous flow of ZU—as is demonstrated clearly on the *Standard Model*. In every instance however:

> *Willingness* defines the range of *Self-Determination*.

SUBJECTIVE WILLINGNESS-PROCESSING

—IMAGINE your physical body is enshrouded in a sphere of light.

—CONSIDER everything you *could know* about *Self*.

—CONSIDER your WILLINGNESS to discover things to *know* about *Self*.

—CONSIDER everything you *could know* about another *Lifeform*.

—CONSIDER your WILLINGNESS to discover things to *know* about *Lifeforms*.

—CONSIDER everything you *could allow* another *Lifeform* to know.

—CONSIDER your WILLINGNESS to assist other *Lifeforms* to discover *Knowingness*.

—CONSIDER everything you *could know* about the *Physical Universe (KI)*.

—CONSIDER your WILLINGNESS to discover things to *know* about the *Physical Universe (KI)*.

—CONSIDER everything you *could Be* in *beta-existence* and as an *Alpha-Spirit*.

—CONSIDER your WILLINGNESS to *Be* these existences in any universe.

—CONSIDER everything you *could allow* other *Lifeforms* to *Be* in *beta-existence* and as an *Alpha-Spirit*.

—CONSIDER your WILLINGNESS to assist other *Lifeforms* to discover *Beingness*.

—CONSIDER everything you *could Do* in *beta-existence* and as an *Alpha-Spirit*.

—CONSIDER your WILLINGNESS to *Do* these actions and sequences in any universe.

—CONSIDER everything you *could allow* other *Lifeforms* to *Do* in *beta-existence* and as an *Alpha-Spirit*.

—CONSIDER your WILLINGNESS to allow other *Lifeforms* to *Do* these things.

—CONSIDER your WILLINGNESS to assist other *Lifeforms* to *Do* these things.

—CONSIDER everything you *could Have* in *beta-existence* and as an *Alpha-Spirit*.

—CONSIDER your WILLINGNESS to *Have* these things in any universe.

—CONSIDER everything you *could allow* other *Lifeforms* to *Have* in *beta-existence* and as an *Alpha-Spirit*.

—CONSIDER your WILLINGNESS to allow other *Lifeforms* to *Have* these things.

—CONSIDER your WILLINGNESS to assist other *Lifeforms* to experience *Havingness*.

—CONSIDER everything you *could change* about your present *Lifetime* experience.

—CONSIDER your WILLINGNESS to *change* these things in your present *Life*.

—CONSIDER everything you *could keep the same* about your present *Lifetime* experience.

—CONSIDER your WILLINGNESS to *keep things from going away* in your present *Life*.

SELF-ACTUALIZING THE WILL "TO BE"

When the *Seeker* can actualize their *Awareness* to a point exterior to *beta-existence*, the first major "seat of consciousness" is WILL, plotted at (5.0) on the ZU-line/*Standard Model*. One can easily see that as a *Master* of WILL—or through *Self-Mastery* of WILL—the individual practices *Willing* to WILL, and thereby actualizes their abilities to *Self-direct* WILL. This is a distinct point of "poise and balance" to *Self-determine* actions in *beta-existence*. But true *Self-Honest* actualization of this point requires *realizations* that take a *Seeker's Awareness* "outside" of—or *exterior* to—*beta-existence*. This includes reducing any point of fragmentation amassed by identifying *Self* too fixedly with the *beta-personality* of the "physical form" and *its* own genetic legacy spreading back countless generations of encoded cellular evolution.

The *ZU-line/Standard Model* was expertly developed and systematized to provide a *map* of reference points between *zero-point continuity* and *Infinity* in the most objective absolute sense. This is necessary for a *Seeker* to be able to orient their "seat of consciousness" in relation to a *consideration* of the ALL. It is easy to become rigidly fixed in certain *Awareness* patterns of expression when an individual is heavily tied down to the *solids* of the Physical Universe, including the *solids* formed by *encoding*, *imprinting* and other reinforced beliefs. As we hold on tightly to these masses, and as more of them accrue, the pendulum of *Awareness* in *beta-experience* swings even harder between extremes or perceived "polar opposites."

An actualized individual is able to observe cause and experience any point on the ZU-line without *becoming* an effect of those points. Also, the individual can only *actualize* points that can be *realized*. We use the *Standard Model* to provide a graphic context to an otherwise *Infinite continuum* of ZU energy. The "seat of consciousness" (*Awareness activity*) is the perspective of *Self* being realized at any given moment, in spite of the fact that the actual *Self*—the "I" or *Alpha-Spirit*—remains in an unchanged state at (7.0) *above* and *beyond* all other "states of *Self*." On the *Standard Model*, we mark *Infinity* as the absolute point beyond that which we can actualize as *Alpha-Spirit* and remain an "I" individuated from the background *Infinity of Nothingness*.

> For a fragmented individual with a limited range of *Awareness*, <u>all</u> points beyond what may be *realized* by the individual are considered *Infinity* (or the "unknown") for *that* individual. For example, an individual with no

sense or *reality* on anything other than the Physical Universe (KI) would then treat all points beyond (0.0) to (4.0) as *Infinity*, because that is the extent of purely *beta-Awareness*.

Individuals (and *Seekers*) who have not yet actualized *Self as* WILL (5.0) often find it difficult to properly divorce the *beta-programming* and *imprints* from the true WILL. This is, by definition, the quality that comes with *Self-Honesty* and *de-fragmentation* of the Human Condition. Beyond WILL (5.0), we approach levels of *Alpha Thought* (6.0) previously undreamed of from lower vantage points—which formerly treated such (now charted) mysteries as an *Unknown Infinity*. So long as the individual *considers* such aspects *to be* outside the scope of the *knowable*, there is no *willingness* to pursue the Path. It is only after we have a sense or demonstration that such things *can be known and realized* in *this* lifetime, that we are *willing* to apply ourselves toward a true journey on the *Pathway to Self-Honesty*—which just as aptly could be considered a part of the greater *Pathway to Infinity*.

When we stretch the reach of our *considerations* toward higher *realizations* beyond the Physical Universe (KI), we are tapping the realm of the "unmanifest"—the "Potential Everythingness" from which all *"thingness"* in *beta-existence* is drawn from. It is a realm of *abstraction*—*motionless* and *cold, unmoved* in *pure intellect* of *Games* and *Logic* that are completely independent of any object of thought and feeling found in *beta-Awareness*. They may be *realized* and even *actualized* into *beta-existence*, but the point where they originate in conception is *itself* independent and *exterior* to the Physical Universe (KI)—and to this point we attribute all *Alpha* (or "Spiritual") existence as-in the direction of "AN" (rendered on the ancient *"Tablets of Destiny"*)—or everything above (4.0) on the *Standard Model*. We can contrast this with the *fragmented* individual that is totally hypnotized with the idea that *they are* the emotional tides and character-roles they assume in *beta-existence*.

EMOTICURVES: MASTERING THE PENDULUM SWING

The emancipated, liberated, actualized *Self* is able to embrace *Awareness* of its own nature—able to dissolve the *illusion* of identifying with the emotional tides, "figure-8" loops[∞] and pendulum swings of reactive extremity—pulling *focus* and *Awareness* this way and that way at the whims of *beta-existence*. The *Actualized Self* becomes a *Master* of *Self* as WILL—a *Master* of the unlimited thought and uninhibited willingness to *express* the *Self* in any manner fitting to the survival and continued existence as *Self* in unity and reach with all *Life* and existence—and to assume these roles as they are seen fit, but to discard them at *Will* lest we linger in any state that is inferior to our *realization* of the true and perfect *Alpha-Spirit* that is never less than "I-AM."

"Cosmic Law" governs *rhythm*, *recursion*[146] and *balance* in the universe—and this is probably demonstrated no better than with the case of *emotional imprints* and *thought fluctuations* that are always present in the face of uncertainty and wherever *Awareness* has been reduced or a Reality has been enforced. All of these various aspects of *beta-existence* fall under the domain of the LAW—and it is the *Mastery* of this Law, not a fear or hatred or apathy toward it, that the mystics and sages of the Ancient Mystery School have ever professed as *The Right Way* to *Ascension*. This way upward on the *Ladder of Lights—Pathway to Infinity*—is a successive reestablishment of the crystal clear *Awareness* of *Self*, and by definition, we are ascending beyond the plane of reactive emotions and conditioned responses that are anything but *Self-determined*. To reach a point of *Ascension* outside or exterior to *beta-existence* is to rise above the pendulum swing of extremity and gain a greater *realization* of the spectrum on which it exists—freely shifting states and changing emotional colors at *Will*.

∞ Ref.—"*Tablets of Destiny*."
146 **recursive** : repeating by looping back onto itself to form continuity; *ex.* the "Infinity" symbol is recursive.

Systemology methods of *psychometeric evaluation* may also be used to demonstrate *trends* and *curves* that indicate "changing states" in the Human Condition. It is only because of "changes" in "states" that we have something to evaluate in relation to the ZU-line/*Standard Model*. The true "I" of *Self* carries no fluctuations or turbulence in its highest actualized position of "I-AM" (7.0). Calculating and graphing a "*Trend of Ascent*" is a useful tool for validating early progress on the "*Pathway to Self-Honesty*" while using *Systemology Processing* and education. When doing so, the *Seeker* is only concerned with plotting the "gains" made toward Ascension—which for basic purposes pertaining to the present volume, is gauged with data established from *Beta-Awareness Tests* (BATs). This is one of the reasons a *Seeker* evaluates such *Self-assessments* at regular intervals or whenever completing a full cycle of work demonstrated in this volume. All *Systemology* "*Emoticurve*"* (√) activity may be easily demonstrated with simple graphs—made from a *square* divided with a "cross" into four equal *squares*.√

As with any graph, the "*Trend of Ascent*" *Emoticurve* is plotted to demonstrate a "rise over a run" or the *amount* of change *over* a period of time. It is only a snapshot on an otherwise infinite continuum and can be used to display any fragment of time—hours, days, weeks, months, &tc. These may be marked along the bottom line of the square representing the period of time being evaluated. If you are still working on the data to plot (you are still planning to add additional plotted points), you can predetermine the total amount of time you intend to graph with a particular *Emoticurve* and write the "start" and "stop" at each end of the bottom line—of course, working to the *right* direction as you progress through time. The "cross" intersects with the bottom timeline at its middle point, which is really an indicator of nothing other than to visually orient the graphic space.

---

* So named for a graph curve based on the "*Emotimeter.*"
√ There is a tendency to use the "square root—√ " sign to indicate *Emoticurves.*

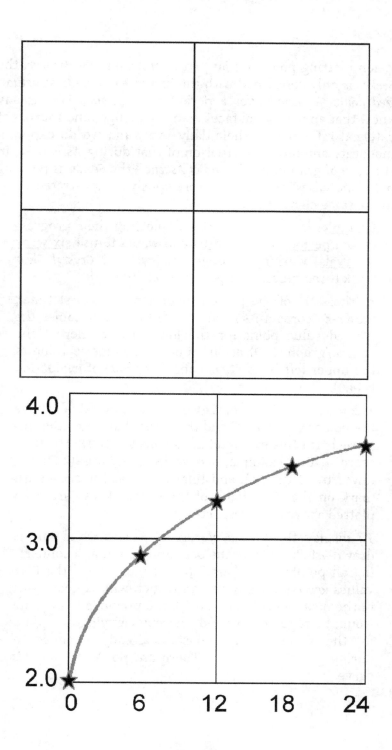

When plotting points on an *"Trend of Ascent"* Emoticurve, the *Seeker* is only concerned with the increases or highest *actualized points* attained over a period of *processing*. We already know that an individual faces many challenges and sources of external influence in their daily lives, and we do expect a moderate amount of fluctuation of that during *Ascent*. So, in the case of graphing a *"Trend of Ascent,"* the *Seeker* is plotting only "increases" to points not previously demonstrated within the same charted time.

> Example: let us say a *Seeker* is plotting their progress over a period of six months and wants to use six week intervals, working through the cycle of *Crystal Clear* work three times, once per six week interval.
>
> At the start of the present book (the very first time), *Seeker-X* scores a BAT value of (1.9). This becomes the plotted value point for the lower left corner of the square/graph: (1.9) at "0" time. The numeric value of the upper left-hand corner for this *Grade* of evaluation is (4.0).
>
> The next point in time, six weeks later, would fall above the point between "0" and the "cross hair line" that intersects (in this example) at "12 weeks" time. The next score in this example, six weeks later, was (2.7); followed by (3.1); (3.4); and finally (3.6) at the six month mark on the right hand of the graph. These are each plotted above their respective times.
>
> To display the *"Trend of Ascent"* Emoticurve, a line is drawn—starting from the beginning point and connecting all points as an "arc." <u>NOTE</u>: Had any of the BAT values *lowered* for any reason (which will be dealt with in the next type of *Emoticurve*), the previous high value would be repeated/plotted, thereby carrying over data for the highest actualized state already ascended to during graphed timeline. There are no "dips" on *this* curve.

During the testing of *Crystal Clear* (prior to its present release), in nearly every instance, the most significant visible increase of "gains" were always found within first lower left square—generally during the first two-to-six weeks (and up to two months) of daily *processing*. This is not to say that additional gains are not achieved—because they are, as is typically demonstrated on your own personal *"Trend of Ascent" Emoticurve*. However, incremental progress at higher levels of *beta-processing* requires targeting increasingly specific *facets* and factors, which tends to require more time and/or deeper applications in order to achieve higher and higher results of equal magnitude. This is actually to be expected, especially regarding *Self-Processing* to the limits presented within this one volume, and should not be interpreted as a flaw in the methods. Given enough time and *Self-determination*—possibly with the assistance of a friend or well-trained *Systemology Pilot*—a fully *Self-dedicated* application of the methodology presented here as the *Crystal Clear Self-Defragmentation Couse* could deliver someone to a basic state of *Self-Honesty* that we have demonstrated as the actualization of a *defragmented* MCC (4.0).

The basic manner of the *Pathway* is demonstrated—albeit crudely portrayed—by the *waveform* of the ZU-line that we interpret using the *Standard Model*. One can see that there is a wider field of range and more potential peaks in the same amount of space as the *frequency increases*. This also implies a bit more "road to travel"—and the *Pathway* winds and bends further on its ascent. If one can imagine divisions of the ZU-line in some way marked on the line—and you were to take a hold of it at each infinity end-point and stretch it out, you would find that there is a considerably greater *quantity* of ZU occupying the "higher" regions than the "lower." We mark them as they are at fixed distances on the *Standard Model* only to demonstrate *relative frequency*—and not actual *spatial distances*. The same situation is found in any depiction of a *kabbalah*, but most specifically "concerning" the steps on the "ancient ziggurat temples"[147] of Babylon, where seven-levels represented the *Seven Spheres* of influence between *Self* and *Infinity* with representations of the planets in their respective orbits. But it was well established that the distances between the planets were not the same and actually exponentially increased the further from the center one was. Today, we can see clearly that the distance, for example between Venus and the Earth is shorter than the distance between Jupiter and Saturn, and so forth.

The other type of *Emoticurve* we are concerned with graphing in this Grade regards very specifically the "descents" and "dips" that are encountered during a *Seeker's* interactions with *beta-existence* combined with their certainty of expressing *Self-determinism* in personal management. For this type of evaluation, we are not concerned with structured scores from a BAT test, and must rely more intuitively on a *Seeker's* (or *Pilot's*) knowledge of the *Emotimeter* and *Awareness Scales*. The same square-and-cross graph-type used above may be applied, except we use the full beta range—(0.0) to (4.0) on the left. Several names for this version have been used at the Offices, but for present purposes, we are calling this one: *"The Pitfall" Emoticurve*.

A true *Self-Honest* examination of the *Pitfalls* in our experience of *beta-existence* actually does require a certain level of *Self-Awareness* to be effectively evaluated by a *Seeker* working alone. For this reason, these graph-types were not demonstrated until this point in the cycle presented in this volume. Although an individual may certain apply this knowledge to examining previous "life-events" on their next *tour* or *cycle* of *Self-Processing* with this book, it is even more critical that the *present* and *future* instances—those taking place after having begun a journey on the *Pathway*—are clearly handled and managed in order to continue making progressive gains. We do not "dip" our *Ascent Emoticurve* because it is in place to es-

---

147 **ziggurat** : ancient Mesopotamian temples in the form of a stepped pyramidal tower presented as a series of seven tiers, levels or terraces.

tablish those points of *actualization* that have already been *attained*. Having traced out that course once, a *Seeker* is more able to *realize* it again—and so the "wide-angle big picture" review of the *Ascent* progress is not penalized at any point. A person may surely taper off and remain at a given point should they decide to do so—and they may even "dip" at times thereafter—but that does not change the point that they *actually did realize* and will ever after carry with them in the back of all *beta-Awareness*.

"*The Pitfall*" *Emoticurve* is used to systematically analyze any moment, interaction or event in *beta-existence* that significantly takes *Self* out of "cause" and puts it in "effect." What we mean by this—for those who have studied *Liber-One*—is the events that disengage *Awareness* from the MCC and engage the RCC. For a *Pitfall Emoticurve*, we are specifically interested in:

a) where an individual *was*: their state of *Awareness* at the beginning or inception of the *Pitfall*;

b) what *sequence of events* took place to result in a *descent* of *beta-Awareness*;

c) the amount of "*dip*" or degree of "differential" between the original (highest) state and the lowest state of *Awareness* succumbed to;

d) the amount of "*time*" spent at any distinguishable state; and

e) the *pathway* taken to get one's *Self-Awareness* out of the *Pitfall* on the other side, after managing (and *processing-out*) any environmental *facets* and *efforts* from others that contributed to this state: including a reduction of any resurfacing/restimulated (past) *imprints* and prevention of any present/future *imprinting* or *fragmentation* as a result of the *Pitfall* event.

Pitfall Emoticurves are managed more efficiently, lessen in their intensity and may disappear altogether as one reaches higher levels of *Awareness* on the *Pathway*. There is no limit to frequencies an individual might *realize* in this *beta-existence* that might render them more and more impenetrable to becoming the *effect* of outside *causes*. This is, in actuality, one of the more idealistic goals behind *Systemology Processing* and the conditions we have established to recognize a *NexGen* rising of *Homo Novus*.

It is of course important that we demonstrate an example of the *Pitfall Emoticurve* before completing this cycle of work. Let us say, for this example, that *Seeker-X* is operating life at their current actualized optimum of (3.5) *Self-confidence*. Suddenly they receive a phone call from a sibling, which now has their attention. Immediately upon engaging at the receiving end of any verbal (or written) communication, *Self-Awareness* drops into a range of beta-attentions or "interest levels" between (3.2) and (2.8). Then, when our sibling delivers the news that our favorite relative has died, *Awareness* immediately descends to "dislike" regarding the present environment. At (2.1) to (2.2) nearly 75% of *beta-Awareness* has been momentarily eliminated—and what happens next regarding Reality dissonance or discordance is a result of the specific individual themselves that is doing the *processing* of information/energy/data in that moment.

Up to this point, we merely have an *event*, not a *Pitfall*. It is not necessary to drop *Awareness* further than this—although the Human Condition is conditioned to "respond" with *emotional reactions*. If this happens—if the *Awareness* drops to (2.0) or *lower*, only then do we have a *Pitfall*—because the RCC has engaged. Otherwise, the information can be managed purely within the level of *Self-determined beta-thought*. This is accomplished by applying *more* personal *Awareness* to *beta-existence* and elevating personal levels back to degrees of "intellectual interest"—thereby maintaining analytical levels of logic and reasoning in our interaction.

Indeed, to the *average person* who knows only a *standard issue* Human Condition, this might seem "cold and unfeeling"—

though interestingly enough those are the same qualities we find of the "truer existence" experienced at the highest levels of the Spiritual Universe (AN). If, however, the RCC is engaging an *emotional reaction*, than the entire *Pitfall* sequence that Humans treat as normal concerning "stages of grief" may all be charted using various degrees on the *Emotimeter* and *Awareness Chart* from (2.0) and *lower*. The solution is already presented as step "e" above.

DEATH AND INITIATION : RECYCLING THE ARTIFICIAL

> "To Bathe in the Waters of New Life;
> To Wash Off the Not-Human;
> —I Come in Self-Annihilation
> and the Grandeur of Inspiration."
> —*William Blake*

Since the inception of the Ancient Mystery School—stretching back on the Earth timeline into prehistory—every esoteric tradition, spiritual system, religious sect, mystical order and secret society has maintained one unique *facet* of "*Self-Transformation*" that has always proven quite effective when properly understood: INITIATION.

For many individuals, the concept of *Initiation* is reduced to the antics of college fraternities that use such *facets* to *imprint* "brotherhood" and "fellowship." It is true that this formal indoctrination into groups takes place—but there is a deeper reason that it *is* effective, even if much of the original symbolism and purpose has since dissolved into obscurity.

—IMAGINE your physical body is enshrouded in a *sphere of light*.

—RECALL a moment you heard someone say the word: *Initiation*.

—RECALL the first time you heard someone say the word: *Initiation*.

—RECALL an instance when you might have seen an *Initiation*.

—RECALL a time when you might have been involved in an *Initiation*.

—ANALYZE every surfacing thought regarding the word: *Initiation*.

Initiation ceremonies are some of the most widely published of all esoteric ceremonies. They shine brightly in contrast to other more abstract philosophies and any underlying work that actually constitutes the bulk of what a *Grade* represents. We can glance back into history, examining ancient Greek and Egyptian philosophical schools and Hermetic priesthoods —not to mention the very blatant demonstration of *seven graded initiation* rituals observed in Mardukite Babylon—and everywhere we find colorful depictions of these in one or another form. In ancient Europe, the Druids would lead initiates into caves—or blindfolded through elaborate labyrinths to further demonstrate the "journey of *Self*" successively reaching to higher and higher points of *Self-Honesty*.

True purpose of INITIATION is to "*Initiate*"—or "start"—a new sequence or cycle of action. Mystical orders and secret society maintain their own methods and terminology to *initiate* members of its tradition, called quite literally "*initiates*." Each point marked by an *Initiation* is both the end of—and start of—a cycle, which denotes that a changed state or "transformation" has observably occurred. No matter what type and flavor is evoked, the meaning behind such "Rites of Passage" remains the same: the death and reformation of the artificial; the shedding of old skin; the rehabilitation of a clearer and greater, more widely encompassing realization of *Self*.

An effective *Initiation Rite* requires more than a sequence of obscure actions and fancy words; it requires more than merely the advancement to some new "title" or "rank" within a social institution; requires far more than merely an ornately decorated setting or lavish temple. In fact, an effective *Initiation Rite* actually requires none of these things mentioned. The most successful *Initiation Rites* simply evoke effective symbolism and "presence" of the most important powerful archetype of *Self-Transformation*:

DEATH OF THE ARTIFICIAL SELF—THE BETA-PERSONALITY.

A FUNERAL FOR THE BETA-PERSONALITY

When originally *initiating* a *Seeker* to various *Grades* of the Ancient Mystery School, a very effective predetermined method would be employed as directed by a particular organization or the *Grade* represented. Each *Grade* completed a cycle of work that cast off another layer of artificial *programming* and inhibiting *personality*—thereby lightening and freeing the *Awareness* of *Self* to realize higher and higher seats of consciousness, moving ever toward the ultimate goal of a perfect reunion of *Awareness* with the *Alpha-Spirit* (7.0). This perfected state of *Self-Actualization* carries its own unique "name" or "title" with each tradition or system—but the ideal is the same, and for once in the history of modern civilization, *Systemology* offers a greater certainty for this achievement than ever realized before!

In keeping with only core symbolism, the method we suggest combines *catharsis* and *kenosis* with physical effort and exertion to create and dissolve a funeral/grave-digging facsimile ceremony.* In essence: you are to dig a grave for your *beta-personality*—all the layers of the "old artificial *Self*" that have been cast off during the previous cycle of work in the present volume. We will not delve much deeper into the theories of initiation at this time, but careful examination of arcane esoteric methods will reveal additional points of logic inherent *behind* the effectiveness of this exercise. For our purposes,

---

* A seeker should be certain they are in appropriate physical health condition before performing such a task.

nearly all *actual* benefit from it occur at the end of a significant cycle of personal *Systemology* work.

Ancient examples of this rite require that an *initiate* in good physical condition essentially dig a physical grave, emptying out a theoretical space for the physical body—yet simultaneously emptying out *Self* of artificial *personality-programming* that has been accumulated; anything that is artificially attributed to the *Identity* that weighs down *Awareness*, fixing it to accept being an effect in *beta-existence*. So, an initiate is instructed to dig in solitude until nearing the point of exhaustion, all the while emptying more and more out from the space associated with the artificial *Identity*. The greater the quantity of artificial attributes; the greater the *efforts* required to cancel out the effects of the *counter-efforts* received during the *Seeker's* lifetime.

Once the grave is dug, the *Seeker* is to contemplate all that has been released from the possession of *Self*—all that has been dissolved and cast off that was only an artificial covering for *Self*, and which was not in any way an actual part of *Self*. The *Seeker* is beside the grave considering all of the *facets* and *aspects* of character and personality, which they had previously assumed and assimilated as *Self*. The greater and more solid this can be imagined as an entire entity separate from *Self*, the more effectively the *Seeker* can imagine that this artificial "body" is laid down into the grave. The *Seeker* will then contemplate that *Self* is no "less" for the fact that it has shed this skin—and is actually able to reach higher points of elation and greater spiritual freedom in return. When this has been successfully imagined, the energies and efforts are recalled by refilling the grave-hole in full comprehension that whatever *was* there is now *dissolved* and the energy and power is *reclaimed* by *Self*.

The *Seeker* has now been reborn to a higher state of *Awareness* and prepared to *initiate* a higher cycle of achievement on the *Pathway to Self-Honesty*.

APPENDIX

GLOSSARY

APPENDIX I

"SYSTEMOLOGY PROCESS AR-SP-2"
[Sumerian Version 2-Step Outline]

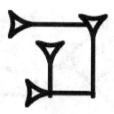

3. **SI** — "to recall; remember; be conscious of in Mind."

Processing Command Line : "Recall ___."

4. **SUG** — "to empty out; to clear; strip away; make naked or bare."

Processing Command Line : "Analyze ___."

APPENDIX II

"SYSTEMOLOGY PROCESS SP-2B"
[Sumerian Version – Command Line]

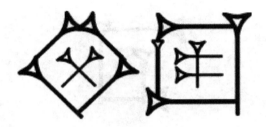

SAG.DAB — "to imagine; conceive of; conjure an idea in Mind."

Processing Command Line : "Imagine ___."

APPENDIX III

"SYSTEMOLOGY PROCESS SP-2B-8A"‡
[ALPHA DEFRAGMENTATION—A.T. 1-8-0∞]

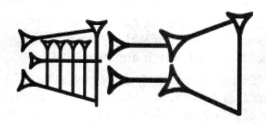

ABZU [ZU.AB] — "Source of All Awareness;
Cosmic Consciousness; Infinity of Nothingness;
Cosmic Ocean; Father of All Beingness."

—IMAGINE your physical body is enshrouded in a *sphere of light*.

—FOCUS your *Awareness* on the *Eighth Sphere* of *Infinity*.

—IMAGINE the *Infinity* of *Nothingness* extending out "infinitely" on all sides as a great Ocean of Cosmic Consciousness.

—FOCUS your *Awareness* from *Self* as a singular focal point of individuated consciousness in the center of the *Infinite Ocean*.

—SENSE that the *Nothingness-Space* all around you is rising up as tides and wave-actions of invisible motion; its abyssal stillness broken by the singular point that is *You*.

---

‡ Supplementing the process described involving Self-realization with the "*8 Spheres & Circles*" from the *Halloween 2019 Extended Course Lecture* (given previously in the present volume)—that process is (unofficially) referred to as "SP-2B-8."

∞ Alpha Defragmentation Procedure. Alpha Technician (AT) Process 1-8-0.

—SENSE that as you press your *Awareness* against the *Nothingness*, there is no resistance, there is no sensation—no feeling of any kind.

—IMAGINE your totality of *Awareness* as the singular focal point of *Infinity*—then REALIZE that the waves you see crashing up against you and rippling into *Infinity* are an extension of your every thought, will and action.

—REALIZE that you are the *Alpha Spirit*; that "wave peak" in an otherwise *Infinity of Nothingness* stretching out within and back off all that was, is and ever WILL.

—REALIZE that your conscious *Awareness* as "I", your direction of WILL as *Alpha Spirit*, and the "central wave peak" born out of *Infinity* are all the same pure individuated ZU—are all *One; None; Infinity*.

—WILL yourself to project *Awareness* ahead of you and see an extension of this ZU as your projection of Identity extending infinitely in front of you—all the way to the *zero-point-continuity* of existence—and back to *Infinity*.

—REPEAT this several times, IMAGINING this ZU as a *Clear Light* radiant extension from *Self*, directed across *Infinity* to *Zero-point* and back to *Infinity*; then REALIZE that you are dissolving and wiping out all *fragmentation* from the channel as you direct the *Clear Light*.

—REPEAT this several times, until you feel confidant in your current results for this cycle of work.

—RECALL the moment you last imagined your physical body enshrouded in a *sphere of light*.

—RECALL the instance you decided to start this present *session*—get a sense of the Intention you *Willed* to begin the session.

—REALIZE that your *beta-Awareness* and the true WILL of the Alpha Spirit are One; End the session.

NEXGEN SYSTEMOLOGY GLOSSARY (ver. 3.2B)

A-for-A : meaning that what we say, write, represent, think or symbolize is a direct and perfect reflection of the actual aspect or thing—that "A" is for, means and is equivalent to "A" and not "a" or "q" or "!"

aberration : a deviation from, or distortion in, what is true or right.

abreaction : fully reliving traumatic past experiences in order to purge them of their emotional excess.

acid-test : an extreme conclusive process to determine the reality, genuineness or truth of a substance or material. This metaphor refers to a process of applying harsh nitric acid to a golden substance (sample) to determine its genuineness.

actualization : to make actual; to bring into Reality; to realize fully in *Awareness*.

affinity : the apparent and energetic *relationship* between substances or bodies; the degree of *attraction* or repulsion between things based on natural forces; the *similitude* of frequencies or waveforms; the degree of *interconnection* between systems.

agreement : unanimity of opinion; an accepted arrangement; "reality."

allegorical : a representation of the abstract, metaphysical or "spiritual" using physical or concrete forms.

alpha : the first, primary, superior or beginning of some form.

alpha control center (ACC) : the highest relay point of *Beingness* for an individuated *Alpha-Spirit, Self* or "I-AM"; in NexGen Systemology—a point of spiritual separation of ZU at (7.0) from the *Infinity of Nothingness* (8.0); the truest actualization of *Identity*; the highest *Self-directed* relay of *Alpha-Self* as an *Identity-Continuum*, operating in an *alpha-existence* (or "Spiritual Universe"–AN) to *determine* "Alpha Thought" (6.0) and WILL-Intention (5.0) *exterior* to the "Physical Universe"–(KI); the "wave-peak" of "I" emerging as individuated consciousness from *Infinity*.

alpha-spirit : a "spiritual" *Life*-form; the "true" *Self* or I-AM; the spiritual (*alpha*) *Self* that is animating the (*beta*) physical body or *"genetic vehicle"* using a continuous *Lifeline* of spiritual (*"ZU"*) energy; an individual spiritual (*alpha*) entity possessing no physical mass or measurable waveform (motion) in the Physical Universe as itself, so it animates the (*beta*) physical body or *"genetic vehicle"* as a catalyst to experience *Self*-determined causality in effect within the *Physical Universe*.

amplitude : the quality of being *ample*; the size or amount of energy that is demonstrated in a *wave*. In the case of audio waves, we associate amplitude with "volume." It is not a statement about the frequencies of waves, only how "loud" they are—to what extent they are or may be projected (or audible).

AN : an ancient cuneiform sign designating the *'spiritual zone'*; the *Spiritual Universe*—comprised of spiritual matter and spiritual energy; a direction of motion toward spiritual *Infinity*, away from or superior to the physical (*'KI'*); the spiritual condition of existence providing for our primary *Alpha* state as an individual *Identity* or *I-AM-Self* which interacts and experiences *Awareness* of a *beta* state in the *Physical Universe* (*'KI'*) as *Life*.

Ancient Mystery School : the original arcane source of all esoteric knowledge.

apotheosis : from the *Greek* word, meaning *"to deify"*; the highest point or apex (for example, of "true knowledge" and "true experience"); an ultimate development of; a glorified or "deified" *ideal*, such as is a quality of *godhood*.

apparent : visibly exposed to sight; evident rather than actual, as presumed by Observation; readily perceived, especially by the senses.

a-priori : from "cause" to "effect"; from a general application to a particular instance; existing in the mind prior to, and independent of experience or observation; validity based on consideration and deduction rather than experience.

archetype : a "first form" or ideal conceptual model of some aspect.

ascension : actualized *Awareness* elevated to the point of true "spiritual existence" exterior to *beta existence*. An "Ascended Master" is one who has returned to an incarnation on Earth as an inherently *Enlightened One*, demonstrable in their actions—they have the ability to *Self-direct* the "Spirit" as *Self*, just as we are treating the "Mind" and "Body" at this current grade of instruction.

assessment scale : an official assignment of graded/gradient numeric values.

assumption : the act of taking or gather to one's Self; taking possession of.

attention : active use of *Awareness* toward a specific aspect or thing.

authoritarian : knowledge as truth, boundaries and freedoms dictated to an individual by a perceived, regulated or enforced "authority."

auto-suggestion (self-hypnosis) : auto-conditioning; self-programming; delivering directed affirmations or statements repeatedly to *Self* in order to condition a change in behavior or beliefs; any *Self-directed* technique intended to generate a specific "*post-hypnotic suggestion.*"

Babylonian : the Mesopotamian civilization that evolved from *Sumer*; the inception of all societal and religious systematization.

band : a division or group; in *NexGen Systemology*—a division or set of frequencies on the ZU-line that are tuned closely together and referred to as a group.

BAT (Beta-Awareness Test) : a method of *psychometric evaluation* developed for *Mardukite Systemology* to determine a "basic" or "average" state of personal *beta-AWARENESS*.

beta (awareness) : all consciousness activity ("*Awareness*") in the "Physical Universe" (KI) or else *beta-existence*; *Awareness* within the range of the *genetic-body*, including material thoughts, emotional responses and physical motors; personal *Awareness* of physical energy and physical matter moving through physical space and experienced as "time"; the *Awareness* held by *Self* that is restricted to a physical organic *Lifeform* or "*genetic vehicle*" in which it experiences causality in the *Physical Universe*.

beta (existence) : all manifestation in the "Physical Universe" (KI); the "Physical" state of existence consisting of vibrations of physical energy and physical matter moving through physical space and experienced as "time"; the conditions of

Awareness for the *Alpha-spirit* (*Self*) as a physical organic *Life-form* or "*genetic vehicle*" in which it experiences causality in the *Physical Universe*.

biological unconsciousness : the organism independent of the sentient *Awareness* of the *Self* to direct it; states induced by severe injury and anesthesia.

cacophony : dissonant, turbulent, harsh and/or discordant sound or noise.

catalog : a systematic list of knowledge or record of data.

catalyst : something that causes action between two systems or aspects, but which itself is unaffected as a variable of this energy communication; a medium or intermediary.

causative : as being the cause.

chakra : (an archaic term used by ancient wisdom traditions); an etheric wheel-mechanism that processes *ZU* energy at specific frequencies along the *ZU-line*, of which a Human being reportedly has *seven* at various degrees.

channel : a specific stream, course, direction or route.

charge : to fill or furnish with a quality; to supply with energy; to lay a command upon; in *NexGen Systemology*—to imbue with intention; to overspread with emotion; application of *Self-directed* "intention" (Will) toward an emotional manifestation in beta-existence.

chronologically : concerning or pertaining to "time."

codification : the process of arranging knowledge in a systematic form.

collapsing a wave : also, "*wave-function collapse*"; the definition or calculation of a wave-function or interaction of potential interactions by Observation. The idea that the Observer is collapsing the wave-function by measuring it appears in the field of *Quantum Physics*. Consciousness or *Awareness* "collapses" the wave-function of energy and matter as the third (required) Principle of Apparent Manifestation.

command line : see "*processing command line.*"

computing device : a calculator or modern computer.

confront : to stand in front of; to meet "face-to-face."

consciousness : the energetic flow of *Awareness*; the Principle System of *Awareness* that is spiritual in nature, which demonstrates potential interaction with all degrees of the Physical Universe; the *Beingness* component of our existence in *Spirit*; the Principle System of *Awareness* as *Spirit* that directs action in the Mind-System.

consideration : careful analytical reflection of all aspects; deliberation; evaluation of facts and importance of certain facts; thorough examination of all aspects related to, or important for, making a decision; the analysis of consequences and estimation of significance when making decisions.

continuity : being a continuous whole; a complete whole or "total round of."

continuum : a continuous *whole*; observing all gradients on a *spectrum*; measuring quantitative variation with gradual transition on a spectrum without demonstrating discontinuity or separate parts.

correlating : the relationship between two or more aspects, parts or systems.

correspondence : a direct relationship or correlation.

Cosmic Law : the "Law" of Nature (or the Physical Universe); the "Law" governing cosmic ordering.

cosmology : a philosophy defining the origins and structure of the universe.

Cosmos : archaic term for the "physical universe"; implies that chaos was brought into order.

counter-productive : contrary to the greater purpose; anything which brings *all Life* away from its sustainable goals of *Infinite Existence*.

crash-coursed : a very intense or steep delivery of education over a very brief time period.

Crossing the Abyss : to enter the spiritual or metaphysical unknown in "Self-annihilation" to purify the Self and "return to the Source."

cuneiform : the oldest extant writing from Mesopotamia; wedge-shaped script inscribed on clay tablets with a reed pen.

cuneiform signs : the cuneiform script, as used in ancient

Mesopotamia, is not represented in a linear alphabet of "letters," but by a systematic use of basic word "signs" that are combined to form more complex word "signs."

data-set : the total accumulation of knowledge used to base Reality.

dead-memories : outdated, inadequate or erroneous data.

defragmentation : the *reparation* of wholeness; a process of removing "*fragmentation*" in data or knowledge to provide a clear understanding; applying techniques and processes that promote a *holistic* interconnected *alpha* state, favoring observational *Awareness* of continuity in all spiritual and physical systems; in *NexGen Systemology*, a "*Seeker*" achieving an actualized state of basic "*Self-Honest Awareness*" is said to be *defragmented*.

degree : a physical or conceptual *unit* (or point) defining the variation present relative to a *scale* above and below it; any stage or extent to which something *is* in relation to other possible positions within a *set* of "*parameters*"; a point within a specific range or spectrum; in *NexGen Systemology*, a *Seeker's* potential energy variations or fluctuations in thought, emotional reaction and physical perception are all treated as "*degrees*."

demographics : identifying segments of the population, real or representative.

destiny : what is set down, made firm, standard, or stands fixed as a constant end; the absolute *destination* regardless of whatever course is traveled; in *NexGen Systemology*, the "*destiny*" of the "*Human Spirit*" (or "*Alpha Spirit*") is infinite existence—"*Immortality*."

dichotomy : a division into two parts, types or kinds.

differential : the quantitative value difference between two forces, motions, pressures or degrees.

differentiation : an apparent difference between aspects or concepts.

discernment : to perceive, distinguish and/or differentiate experience into true knowledge.

displace : to compel to leave; to move or replace something with something else in its place or space.

dissonance : discordance; out of step; out of phase; disharmonious; the "differential" between the way things are and the way things are experienced.

dross : prime material; specifically waste-matter or refuse; the discarded remains collected together.

dynamic (systems) : a principle or fixed system which demonstrates its *'variations'* in activity (or output) only in constant relation to variables or fluctuation of interrelated systems; a standard principle, function, process or system that exhibits *'variations'* and change simultaneously with all connected systems.

Eastern traditions : the evolution of the *Ancient Mystery School* east of its origins, primarily the Asian continent, or what is archaically referred to as "oriental."

echelon : a level or rung on a ladder; a rank or level of command.

eclipse : to cast a shadow or darken; to block out or obscure a comparison.

emotional encoding : the substance of *imprints*; associations of sensory experience with an *imprint*; perceptions of our environment that receive an *emotional charge*, which form or reinforce facets of an *imprint*; perceptions recorded and stored as an *imprint* within the "emotional range" of energetic manifestation.

enact : to make happen; to bring into action; to make part of an act.

encompassing : to form a circle around, surround or envelop around.

energy signatures : a distinctive pattern of energetic action.

enforcement : the act of compelling or putting (effort) into force; to compel or impose obedience by force; to impress strongly with applications of stress to demand agreement or validation.

engineering : the *Self-directed* actions and efforts to utilize knowledge (observed causality/science), maths (calculations/quantification) and logic (axioms/formulas) to understand, design or manifest a solid structure, machine, mechanism, engine or system; as *"Reality Engineering"* in *Nex-*

Gen Systemology—intentional *Self-directed* adjustment of existing Reality conditions; the application of total *Self-determinism* in *Self-Honesty* to change apparent Reality using fundamentals of *Systemology* and *Cosmic Law*.

entanglement : tangled together; intertwined and enmeshed systems; in *NexGen Systemology*, a reference to the interrelation of all particles as waves at a higher point of connectivity than is apparent, since wave-functions only "collapse" when someone is *Observing*, or doing the measuring, evaluating, &tc.

entropy : the reduction of organized physical systems back into chaos-continuity when their integrity is measured against space over time.

epicenter : the point from which shock-waves travel.

epistemology : a school of philosophy focused on the truth of knowledge *and* knowledge of truth; theories regarding validity and truth inherent in any structure of knowledge and reason.

erroneous : inaccurate; incorrect; containing error.

esoteric : hidden; secret; knowledge understood by a select few.

etching : to cut, bite or corrode with acid to produce a pattern.

evaluate : to determine, assign or fix a set value, amount or meaning.

exacting : a demanding rigid effort to draw forth from.

executable : the supreme authoritative ability to carry out according to design.

existence : the *state* or fact of *apparent manifestation*; the resulting combination of the Principles of Manifestation: consciousness, motion and substance; continued *survival*; that which independently exists; the *'Prime Directive'* and sole purpose of all manifestation or Reality; the highest common intended motivation driving any "*Thing*" or *Life*.

existential : pertaining to existence, or some aspect or condition of existence.

extant : in existence; existing.

exoteric : public knowledge or common understanding; the level of understanding and *Knowing* maintained by the "masses"; the opposite of *esoteric*.

experiential data : accumulated reference points we store as memory concerning our "experience" with Reality.

extrapolate : to make an estimate of the "value" outside of the perceivable range.

extropy : in *NexGen Systemology*—the reduction of organized spiritual systems back into a singularity of Infinity when their integrity is measured against space over time.

facets : an aspect, an apparent phase; one of many faces of something; a cut surface on a gem or crystal; in *NexGen Systemology*—a single perception or aspect of a memory or "*Imprint*"; any one of many ways in which a memory is recorded; perceptions associated with a painful emotional (sensation) experience and "*imprinted*" onto a metaphoric lens through which to view future similar experiences.

faculties : abilities of the mind (individual) inherent or developed.

fallacy : a deceptive, misleading, erroneous and/or false beliefs; unsound logic; persuasions, invalidation or enforcement of Reality agreements based on authority, sympathy, bandwagon/mob mentality, vanity, ambiguity, suppression of information, and/or presentation of false dichotomies.

fate : what is brought to light or actualized as experience; the actual *course* taken to reach an end, charted end, or final *destination*; in *NexGen Systemology*, the '*fate*' of a '*Human Spirit*' (or '*Alpha Spirit*') is determined by the choice of course taken to experience *Life*.

feedback loop : a complete and continuous circuit flow of energy or information directed as an output from a source to a target which is altered and return back to the source as an input; in *General Systemology*—the continuous process where outputs of a system are routed back as inputs to complete a circuit or loop, which may be closed or connected to other systems/circuits; in *NexGen Systemology*—the continuous process where directed *Life* energy and *Awareness* is sent back to *Self* as experience, understanding and memory to complete an energetic circuit as a loop.

flattening a wave : see "*process-out.*"

fragmentation : breaking into parts and scattering the pieces; the *fractioning* of wholeness or the *fracture* of a holistic interconnected *alpha* state, favoring observational *Awareness* of perceived connectivity between parts; *discontinuity*; separation of a totality into parts; in *NexGen Systemology*, a person outside a state of *Self-Honesty* is said to be *fragmented*.

game theory : a mathematical theory of logic pertaining to strategies of maximizing gains and minimizing loses within prescribed boundaries and freedoms; a field of knowledge widely applied to human problem solving and decision-making.

general systemology ("systematology") : a methodology of analysis and evaluation regarding the systems—their design and function; organizing systems of interrelated information-processing in order to perform a given function or pattern of functions.

genetic memory : the evolutionary, cellular and genetic (DNA) "memory" encoded into a *genetic vehicle* or *living organism* during its progression and duplication (reproduction) over millions (or billions) of years on Earth; in *NexGen Systemology*— the past-life Earth-memory carried in the genetic makeup of an organism (*genetic vehicle*) that is *independent of any* actual "spiritual memory" maintained by the *Alpha Spirit* themselves, from its own previous lifetimes on Earth and elsewhere using other *genetic vehicles* with no direct evolutionary connection to the current physical form in use.

genetic-vehicle : a physical *Life*-form; the physical (*beta*) body that is animated/controlled by the (*Alpha*) *Spirit* using a continuous *Lifeline* (ZU); a physical (*beta*) organic receptacle and catalyst for the (*Alpha*) *Self* to operate "causes" and experience "effects" within the *Physical Universe*.

gifted : attributing a special quality or ability; having exceptionally high intelligence or mental faculties.

gnosis : a *Greek* word meaning knowledge, but specifically "true knowledge"; the highest echelon of "true knowledge" accessible (or attained) only by mystical or spiritual faculties.

godhood : a divine character or condition; "divinity."

gradient : a degree of partitioned ascent or descent along some scale, elevation or incline; "higher" and "lower" values in relation to one another.

heralded : proclaimed ahead of or prior to; officially announced.

holistic : the examination of interconnected systems as encompassing something greater than the *sum* of their "parts."

Homo Novus : literally, the "new man"; the "newly elevated man" or "known man" in ancient Rome; the man who "knows (only) through himself"; in NexGen Systemology—the next spiritual and intellectual evolution of *homo sapiens* (the "modern Human Condition"), which is signified by a demonstration of higher faculties of *Self-Actualization* and clear *Awareness*.

Homo Sapiens Sapiens : the present standard-issue Human Condition; the *hominid* species and genetic-line on Earth that received modification, programming and conditioning by the *Anunnaki* race of *Alpha-Spirits*, of which early alterations contributed to various upgrades (changes) to the genetic-line, beginning approximately 450,000 years ago (*ya*) when the *Anunnaki* first appear on Earth; a species for the Human Condition on Earth that resulted from many specific *Anunnaki* "genetic" and "cultural" *interventions* at certain points of significant advancement—specifically *circa* 300,000 *ya*, 200,000 *ya*, 40,000 *ya*, and 8,000 *ya*; a species of the Human Condition set for replacement by *Homo Novus*.

Human Condition : a standard default state of Human experience.

humanistic psychology : a field of academic psychology approaching a holistic emphasis on *Self-Actualization* as an individual's most basic motivation; early key figures from the 20th century include: Carl Rogers, Abraham Maslow, L. Ron Hubbard, William Walker Atkinson, Deepak Chopra and Timothy Leary (to name a few).

hypothetical : operating under the assumption a certain aspect actual "is."

identity : the collection of energy and matter—including memory—across the "*Spiritual Continuum*" that we consider as "I" of *Self*.

identity-system : the application of the *ZU-line* as "I"—the continuous expression of *Self* as *Awareness*.

illuminated : to supply with light so as to make visible or comprehensible.

immersion : plunged or sunk into; wholly surrounded by.

imprint : to strongly impress, stamp, mark (or outline) onto a softer 'impressible' substance; to mark with pressure onto a surface; in *NexGen Systemology*, the term is used to indicate permanent Reality impressions marked by frequencies, energies or interactions experienced during periods of emotional distress, pain, unconsciousness or antagonism to physical survival, all of which are are stored with other reactive response-mechanisms at lower-levels of *Awareness* as opposed to the active memory database and proactive processing center of the Mind; an experiential "memory-set" that may later resurface—be triggered or stimulated artificially—as Reality, of which similar responses will be engaged automatically.

incarnation : a present, living or concrete form of some thing or idea.

inception : the beginning, start, origin or outset.

incite : to urge on; instigate; prove or stimulate into action.

indefinable : without a clear definition being currently presented.

individual : a person, human entity or *Seeker* is sometimes referred to as an "individual" within this text.

infinite existence : "immortality."

inhibited : withheld, discouraged or repressed from some state.

"in phase" : see *"phase alignment."*

institution : a social standard or organizational group responsible for promoting some system or aspect in society.

interdimensional : systems that are interconnected or correlated between the Physical Universe and the Spiritual Universe —or between "dimension states" observably identified as "physical," "emotional," "psychological" and "spiritual." The only point of true interconnectivity that we can systematically determine is called *"Life."*

intermediate : a distinct point between two points; actions between two points.

invalidate : decrease the level or degree or *agreement* as Reality.

invests : spends on; gives or devotes something to earn a result; endows with.

knowledge : clear personal processing of informed understanding; information (data) that is actualized as effectively workable understanding; a demonstrable understanding on which we may 'set' our *Awareness*—or literally a "know-ledge."

KI : an ancient cuneiform sign designating the *'physical zone'*; the *Physical Universe*—comprised of physical matter and physical energy in action across space and observed as time; a direction of motion toward material *Continuity*, away from or subordinate to the Spiritual (*'AN'*); the physical condition of existence providing for our *beta* state of *Awareness* experienced (and interacted with) as an individual *Lifeform* from our primary Alpha state of Identity or *I-AM-Self* in the *Spiritual Universe* (*'AN'*).

kinetic : pertaining to the energy of physical motion and movement.

learned : highly educated; possessing significant knowledge.

level : a physical or conceptual *tier* (or plane) relative to a *scale* above and below it; a significant *gradient* observable as a *foundation* (or surface) built upon and subsequent to other levels of a totality or whole; a *set* of "*parameters*" with respect to other such *sets* along a *continuum*; in NexGen Systemology, a *Seeker's* understanding, *Awareness* as *Self* and the formal grades of material/instruction are all treated as "*levels*."

localized : brought together and confined to a particular place.

logic equations : using symbols and basic mathematical logic to establish the validity of statements or to see how a variable within a system will change the result; a basic demonstration of proportion or relationship between variables in a system.

logistics : pertaining to the movement or transportation between locations.

macrocosmic : taking examples and system demonstrations at one level and applying them as a larger demonstration of a relatively higher level or unseen dimension.

manifestation : something brought into existence.

Marduk : founder of Babylonia; patron Anunnaki "god" of Babylon.

Mardukite Zuism : a Mesopotamian-themed (Babylonian-oriented) religious philosophy and tradition applying the spiritual technology based on *Arcane Tablets* in combination the "Tech" from *NexGen Systemology*; first developed in the New Age underground by Joshua Free in 2008 and realized publicly in 2009 with the formal establishment of the *"Mardukite Chamberlains."*

master control center (MCC) : a perfect computing device to the extent of the information received from "lower levels" of sensory experience/perception; the proactive communication system of the *"Mind"*; a relay point of active *Awareness* along the Identity's *ZU-line*, which is responsible for maintaining basic *Self-Honest Clarity* of *Knowingness* as a *seat of consciousness* between the *Alpha-Spirit* and the secondary *"Reactive Control Center"* of a *Lifeform* in *beta existence*; the Mind-center for an *Alpha-Spirit* to actualize cause in the *beta existence*; the analytical *Self-Determined* Mind-center of an *Alpha-Spirit used* to project *Will* toward the genetic body; the point of contact between *Spiritual Systems* and the *beta existence*; presumably the *"Third Eye"* of a being connected directly to the *I-AM-Self*, which is responsible for *determining* Reality at any time; in *NexGen Systemology*, this is plotted at (4.0) on the continuity model of the *ZU-line*.

Mesopotamia : land between Tigris and Euphrates River; modern-day Iraq.

methodology : a system of methods, principles and rules to compose a systematic paradigm of philosophy or science.

"Mind's Eye" : the activities of the "Third-Eye" (or actualized MCC) where the *Alpha-Spirit* directly interacts with the organic *genetic vehicle* in *beta-existence*; *Self-directed* activity on the plane of "mental consciousness" that is maintained between "spiritual consciousness" of the *Alpha-Spirit* and the "physical/emotional consciousness" of the *genetic vehicle*; the "consciousness activity" *Self-directed* by an actualized WILL.

misappropriated : put into use incorrectly; to apply ineffectively or as unintended by design.

motor functions : internal mechanisms that allow a body to move.

Nabu : the original "god of wisdom, writing and knowledge." (Babylonian)

negligible : so small or trifle that it may be disregarded.

NexGen Systemology : a method of applied religious philosophy and spiritual technology based on *Arcane Tablets* in combination with *"general systemology"* and *"games theory"* developed in the New Age underground by Joshua Free in 2011 as an advanced futurist extension of the *"Mardukite Chamberlains."*

objectively : concerning the "external world" and attempts to observe Reality independent of personal "subjective" factors.

optimum : the most favorable conditions for the best result; the great degree of result under specific conditions.

orchestration : to arrange or compose the performance of a system.

organic : as related to a physically living organism or carbon-based life form.

oscillation-alternation : a particular type of (or fluctuation) between two relative states, conditions or degrees; a wave-action between two degrees, such as is described in the action of the *pendulum effect*; a flux or wave-like energy in motion, across space, calculable as time.

pantheism : religious philosophies that observe God as inherent within all aspects of the Physical Universe.

paradigm : an all-encompassing *standard* by which to view the world and *communicate* Reality; a standard model of reality-systems used by the Mind to filter, organize and interpret experience of Reality.

parameters : a defined range of possible variables within a model, spectrum or continuum.

paramount : the most important; "above all else."

participation : being part of the action or affecting the result.

patron god : the most sacred deity of a region or city, of which most temples and religious services are directed; the personal deity of an individual.

perturbation : the deviation from a natural state, fixed motion, or orbit system caused by another external system; disturbing or disquieting the serenity of an existent state; inciting observable apparent action using indirect or outside actions or 'forces'; the introduction of a new element or facet that disturbs equilibrium of a standard system; the "butterfly effect"; in *NexGen Systemology*, *'perturbation'* is a necessary condition for the *ZU-line* to function as a *Standard Model* of actual *'monistic continuity'*—which is a *Lifeforce* singularity expressed along a spectrum with potential interactions at each degree from any source; the influence of a degree in one state by activities of another state that seem independent, but which are actually connected directly at some higher degree, even if not apparently observed.

phase alignment or *"in phase"* : to be in synch, in step or aligned properly with something else in order to increase the total strength value; in *NexGen Systemology*—referring to alignment of *Awareness* with a particular identity, space or time (such as being *in Self in* present *space* and *time*).

philanthropy : charitable; the intention (or programmed desire) to generously provide personal wealth and service to the well-being and continued existence of others.

physics : a science of motions, forces and bodies in the Physical Universe.

physiology : a science of motions of living bodies or organisms.

pilfering : to steal in small quantities; petty theft.

pilot : the steersman of a ship; in *NexGen Systemology*—an individual qualified to operate *Systemology Processing* for other *Seekers* on the *Pathway to Self-Honesty*.

postulate : to put forward as truth; to suggest or assume an existence *to be*; to provide a basis of reasoning and belief; a basic theory accepted as fact.

precedent : a matter which precedes or goes before another in importance.

precipitate : to actively hasten or quicken into existence.

prehistoric : any time before human history is written; prior to c. 4000 B.C.

premise : a basis or statement of fact from which conclusions are drawn.

prevalent : of wide extent; an extensive or largely accepted aspect or current state.

"process-out" or **"flatten a wave"** : to reduce *emotional encoding* of an *imprint* to zero; to dissolve a *wave-form* or *thought-formed* "solid" such as a *"belief"*; to completely run a *process* to its end, thereby *flattening* any previously *"collapsed-waves"* or *fragmentation* that is obstructing the *clear channel* of *Self-Awareness*; also referred to as "processing-out."

processing command line (PCL) or **command line** : a directed input; a specific command using highly selective language for *Systemology Processing*; a predetermined directive statement intended to focus concentrated attention.

projecting awareness : sending out (motion) or radiating *"consciousness"* from *Self* ("I").

proportional : having a direct relationship or mutual interaction with.

Proto-Indo-European (PIE) : a single source root language c.4500 B.C. contributing to most European languages.

psychometric evaluation : the relative measurement of personal ability, mental (psychological/thought) faculties, and effective processing of information and external stimulus data; a scale used in "applied psychology" to evaluate and predict human behavior.

reactive control center (RCC) : the secondary (reactive) communication system of the *"Mind"*; a relay point of *Awareness* along the Identity's *ZU-line*, which is responsible for engaging basic motors, biochemical processes and any *programmed automated responses* of a living *beta* organism; the reactive Mind-Center of a living organism relaying communications of *Awareness* between causal experience of *Physical Systems* and the *"Master Control Center"*; it presumably stores all emotional encoded imprints as fragmentation of "chakra" frequencies of *ZU* (within the range of the *"psychological/emotive systems"* of a being), which it may *react* to as Reality at any time; in *NexGen Systemology*, this is plotted at (2.0) on the continuity model of the *ZU-line*.

receptacles : a device or mechanism designed to contain and store a specific type of aspect or thing; a container meant to receive something.

recursive : repeating by looping back onto itself to form continuity; *ex.* the "Infinity" symbol is recursive.

relative : an apparent point, state or condition that treated distinct from others.

relinquish : to give up control, command or possession of.

repetitively : to repeat "over and over" again; or else "repetition."

responsibility : the *ability* to *respond*; the extent of mobilizing *power* and *understanding* an individual maintains as *Awareness* to enact *change*; the proactive ability to *Self-direct* and make decisions independent of an outside authority.

resurface : to return to, or bring up to, the "surface" what has been submerged; in *NexGen Systemology*—relating specifically to processes where a *Seeker* recalls blocked energy stored covertly as emotional "*imprints*" (by the RCC) so that it may be effectively defragmented from the "*ZU-line*" (by the MCC).

scions : a descendant or child offspring; an offshoot or branch.

Seeker : an individual on the *Pathway to Self-Honesty*; a practitioner of *Mardukite Systemology* or *NexGen Systemology Processing* that is working toward *Ascension*.

Self-actualization : bringing the full potential of the Human spirit into Reality; expressing full capabilities and creativeness of the *Alpha-Spirit*.

Self-determinism : the freedom to act, clear of external control or influence.

Self-evaluation : see "*psychometric evaluation.*"

Self-honesty : the *alpha* state of *being* and *knowing*; clear and present total *Awareness* of-and-as *Self*, in its most basic and true proactive expression of itself as *Spirit* or *I-AM*—free of artificial attachments, perceptive filters and other emotionally-reactive or mentally-conditioned programming imposed on the human condition by the systematized physical world.

self-sustained : self-supported.

semantics : the *meaning* carried in *language* as the *truth* of a "thing" represented, A-for-A; the *effect* of language on *thought* activity in the Mind and physical behavior; language as *symbols* used to represent a concept, "thing" or "solid."

semantic-set : the implied meaning behind any words or symbols used in a specific paradigm.

sentient : consciously intelligent.

simulacrum : an tangible image, facsimile or superficial representation that carries a likeness or similarity to someone or something else; in *NexGen Systemology*—the *genetic vehicle* or physical body is an example of a "simulacrum" of the true *Alpha-Spirit* or *Self* (I-AM), which otherwise has no tangible form in *beta-existence*.

sine-wave : the *frequency* and amplitude of a quantified (calculable) *vibration* represented on a graph (graphically) as smooth repetitive *oscillation* of a *waveform*; a *waveform* graphed for demonstration—otherwise represented in *NexGen Systemology* logic equations as 'Wf,' or in mathematics as the '*function of x*' (fx); graphically representing arcs (*parameters*) of a circular *continuity* on a *continuum*; in the *Standard Model of NexGen Systemology*, the actual 'wave vibration' graphically displayed on an otherwise static *ZU-line* (of Infinity) is a '*sine-wave*'.

singularity : a point where apparently dissimilar qualities of all aspects share a singular expression, nature or quality.

slate : a flat surface used for writing on; a chalk-board.

somatic : pertaining to the physical body and its response actions.

spectrum : a broad range or array as a continuous series or sequence.

standard issue : equally dispensed to all without consideration.

standard model : a fundamental *structure* or symbolic construct used to evaluate a complete *set* in *continuity* relative to itself and variable to all other *dynamic systems* as graphed or calculated by *logic*; in *NexGen Systemology*—a "*monistic continuity model*" demonstrating *total system* interconnectivity "above" and "below" observation of any apparent *parameters*; the *ZU-line* represented as a singular vertical (*y*-axis) waveform in space without charting any specific movement across a dimensional time-graph *x*-axis.

static : characterized by a fixed or stationary condition; having no apparent change, movement or fluctuation.

stoicism : pertaining to the school of "stoic" philosophy, distinguished by calm mental attitudes, freedom from desire/passion and essentially any emotional fluctuation.

sub-zones : at ranges "below" which we are representing or which is readily observable for current purposes.

successively : what comes after; forward into the future.

succumb : to give way, or give in to, a relatively stronger superior force.

Sumerian : ancient civilization of *Sumer*, founded in Mesopotamia c. 5000 B.C.

superfluous : excessive; unnecessary; needless.

superstition : knowledge accepted without good reason.

surefooted : proceeding surely; not likely to stumble or fall.

symbiotic : pertaining to the closeness, proximity and affinity between two beings that are in mutual communication or maintaining mutually validating interactions.

sympathy : a sensation, feeling or emotion—of anger, fear, sorrow and/or pity—that is a *personal reaction* to the misfortune and failure of another being.

systematization : to arrange into systems; to systematize or make systematic.

thought-experiment : from the German, *Gedankenexperiment*; logical *considerations* or mental models used to concisely visualize consequences (cause-effect sequences) within the context of an imaginary or hypothetical scenario; using faculties of the Mind's Eye to *Imagine* things accurately with *considerations* that *have not* already been consciously experienced in *beta-existence*.

thought-form : apparent *manifestation* or existential *realization* of Thought-waves as "solids" even when only apparent in Reality-agreements of the Observer; the treatment of *Thought-waves* as permanent *imprints* obscuring *Self-Honest Clarity* of *Awareness* when reinforced by emotional experience as actualized "thought-formed solids" ("*beliefs*") in the Mind.

thought-habit : reoccurring modes of thought or repeated "self-talk"; essentially "self-hypnosis" resulting in a certain state.

thought-wave : a proactive *Self-directed action* or reactive-response *action* of *consciousness*; the *process* of *thinking* as demonstrated in *wave-form*; the *activity* of *Awareness* within the range of *thought vibrations/frequencies* on the existential *Life-continuum* or *ZU-line*.

threshold : a doorway, gate or entrance point; the degree to which something is to produce an effect within a certain state or condition.

thwarted : to successfully oppose or prevent a purpose from actualizing.

tier : a series of rows or levels, one stacked immediately before or atop another.

timeline : plotting out history in a linear (line) modal to indicated instances (experiences) or demonstrate changes in state (space) as measured over time.

tipping point : a definitive "point" when a series of small changes (to a system) are significant enough to be *realized* or *cause* a larger, more significant change; the critical "point" (in a system) beyond which significant change takes place or is observed; the "point" at which changes that cross a specific "threshold" reach a noticeably new state or development.

transhumanism : concerning the next evolved state of the "Human Condition," which is to say either in a direction of "internal" or "spiritual" technologies that advance the *Self*, or the direction of "external" and "physical" technologies that either modify or eliminate the *Body*. In our present state of society, it is the "physical" that is selectively *sold* to the masses so that only a select few may experience the "former."

transmit : to send forth data along some line of communication.

traumatic encoding : information received when the sensory faculties of an organism are "shocked" into learning it as an "emotionally" encoded *"Imprint."*

unconscious : a state when *Awareness* as *Self* is removed from the equation of *Life* experience.

undefiled : to remain intact, untouched or unchanged; to be left in an original "virgin" state.

validation : the reinforcement of agreements of Reality.

vantage : a point, place or position that offers a good view.

Venn diagram : a diagram for symbolic logic using circles to represent sets and their systematic relationship; named after the logician *John Venn*.

verbatim : precisely reproduced "word" for "word."

vibration : effects of motion or wave-frequency as applied to any system.

vizier : a high ranking official; a minister-of-state.

wave-function collapse : see *"collapsing a wave."*

Western Civilization : the modern history, culture, ideals, values and technology, particularly of Europe and North America as distinguished by growing urbanization and industrialization.

will *or* **WILL** (5.0) : in *NexGen Systemology* (from the *Standard Model*)—the spiritual ability at (5.0) of an *Alpha Spirit* (7.0) to apply *intention* as "Cause" from a higher order of reasoning and consideration (6.0) than the thoughts found in *beta-existence*, where it manifests as "effect" below (4.0).

ziggurat : ancient Mesopotamian temples in the form of a stepped pyramidal tower presented as a series of seven tiers, levels or terraces.

ZU : the ancient cuneiform sign designating an archaic verb —*"to know," "knowingness"* or *"awareness"*; the active energy/matter of the "Spiritual Universe" (AN) that is experienced as *Lifeforce* or *consciousness* for entities existing in the "Physical Universe" (KI); *"Spiritual Life Energy"*; the spiritual energy present in the WILL of the actualized *Alpha-Spirit* in the "Spiritual Universe" (AN), which imbues its *Awareness* into the Physical Universe (KI), animating/controlling *Life* for its experience of *beta-existence* along an *Identity-continuum* called a *ZU-line*.

ZU-line : a spectrum of *Spiritual Life Energy*; an energetic channel of *Identity-continuum* connecting the *Awareness* (ZU) of an *Alpha-Spirit* with *"Infinity"*; a *Life-line* on which *Awareness* (ZU) extends from the direction of the "Spiritual Universe" (AN) as its *alpha state* through an entire possible range of activity in its *beta state*, experienced as a *genetic-entity* occupying the *Physical Universe (KI)*.

WOULD YOU LIKE TO KNOW MORE ???

Your
Pathway to Self-Honesty
continues with books and courses by
Joshua Free

<u>NEW MATERIALS COMING IN THE 2020's</u>

from
the NexGen Systemological Society (NSS)
and
International School of Systemology (ISS)
a division of the
Mardukite Research Organization

stay tuned!

SYSTEMOLOGY
The Pathway to Self-Honesty

THE TABLETS OF DESTINY
*Using Ancient Wisdom
to Unlock Human Potential*
by Joshua Free

(Mardukite Systemology Liber-One)

A rediscovery of the original
System of perfecting the Human
Condition on a Pathway which
leads to Infinity.
Here is a new map on which
to chart the future
spiritual evolution of
all humanity!

**SYSTEMOLOGY:
THE ORIGINAL THESIS**
*An Introduction to
21st Century New Thought*
by Joshua Free

(Mardukite Systemology Liber-S-1X)

A collection of the original
underground discourses released to
the "New Thought" division of the
Mardukite Research Organization
and providing the inspiration for
rapid futurist spiritual technology
called "Mardukite Systemology."

- - - - - - - - - - - - - - - - -ALSO AVAILABLE- - - - - - - - - - - - - - - - -

• Systemology: Truth Seeker's Adventure Journal & Flight Log
• The Power of Zu: Controlling the Radiant Energy *(Liber S-1Z)*
• Mardukite Zuism: A Brief Introduction *(Liber S-1Y)*
• Communication and Control of Energy & Power *(Liber 2C)*

SYSTEMOLOGY
The Pathway to Self-Honesty

The ultimate operator's manual to the Human Condition and unlocking the true power of the Alpha Spirit

"Modern Mardukite Zuism"
"The Tablets of Destiny"
"Crystal Clear"
"Systemology: The Original Thesis"
—Human, More Than Human
—Defragmentation
—Patterns & Cycles
—Transhuman Generations
"The Power of ZU"
"Grade-III Master Course"
—Bonus Material

THE COMPLETE GRADE-III COLLECTION IN ONE VOUME !!!

THE SYSTEMOLOGY HANDBOOK
MASTER EDITION HARDCOVER BY JOSHUA FREE

A decade of collected writings revealing the full research and discovery of new advancements in 21st century "New Thought" are culminated together to present the most complete guide, reference and course curriculum of "Mardukite Systemology" established to date.

Here is the technology to ensure a true spiritual "transhuman" evolution into the future. Underground materials originally composing over 8 books in total are expertly arranged for Truth Seekers in one amazing volume!

For the first time, anyone can gain unhindered access to the most complete collection of practical "NexGen" teachings and techniques drawn directly from the secret knowledge and wisdom of Arcane Tablets of the Ancient Mystery School. We are standing at the cusp of a true "New Age" for humanity—standing in witness to the dawning light on the horizon of a Crystal Age.

EXISTING MARDUKITE RESEARCH LIBRARY ARCHIVE TITLES
AVAILABLE FROM THE **JOSHUA FREE** PUBLISHING IMPRINT

Necronomicon: The Anunnaki Bible : 10th Anniversary
  Collector's Edition—LIBER-N,L,G,9+W-M+S *(Hardcover)*

The Complete Anunnaki Bible: A Source Book of Esoteric Archaeology :
  10th Anniversary—LIBER-N,L,G,9+W-M+S *(Paperback)*

Necronomicon: The Anunnaki Bible : 10th Anniversary
  Pocket Edition—*(Abridged Paperback)*

*Gates of the Necronomicon: The Secret Anunnaki Tradition of
  Babylon* : 10th Anniversary Collector's Edition—
    LIBER-50,51/52,R+555 *(Hardcover)*

The Sumerian Legacy: A Guide to Esoteric Archaeology—
    LIBER-50+51/52 *(Paperback)*

*Necronomicon Revelations—Crossing to the Abyss: Nine Gates
  of the Kingdom of Shadows & Simon's Necronomicon*—
    LIBER-R+555 *(Paperback)*

*Necronomicon: The Anunnaki Grimoire: A Manual of Practical
  Babylonian Magick* : 10th Anniversary Collector's Edition—
    LIBER-E,W/Z,M+K *(Hardcover)*

*Practical Babylonian Magic : Invoking the Power of the Sumerian
  Anunnaki*—LIBER-E,W/Z,M+K *(Paperback)*

*The Complete Book of Marduk by Nabu : A Pocket Anunnaki
  Devotional Companion to Babylonian Prayers & Rituals* :
    10th Anniversary Collector's Edition—LIBER-W+Z *(Hardcover)*

*The Maqlu Ritual Book : A Pocket Companion to Babylonian
  Exorcisms, Banishing Rites & Protective Spells* :
    10th Anniversary Collector's Edition—LIBER-M *(Hardcover)*

Necronomicon: The Anunnaki Spellbook : 10th Anniversary
  Pocket Edition—LIBER-W/Z+M *(Abridged Paperback)*

*The Anunnaki Tarot : Consulting the Babylonian Oracle of
  Cosmic Wisdom (Guide Book)*—LIBER-T *(Paperback)*

*Elvenomicon—or—Secret Traditions of Elves & Faeries : The Book of
  Elven Magick & Druid Lore :* 15th Anniversary Collector's
    Edition—LIBER-D *(Hardcover)*

The Druid's Handbook : Ancient Magick for a New Age
    20th Anniversary Collector's Edition—LIBER-D2 *(Hardcover)*

Draconomicon : The Book of Ancient Dragon Magick :
    25th Anniversary Collector's Edition—LIBER-D3 *(Hardcover)*

The Sorcerer's Handbook : A Complete Guide to Practical Magick
    21st Anniversary Collector's Edition—*(Hardcover)*

∞

"I find myself in the unique position of <u>piloting</u> the course of humanity's future survival and spiritual evolution in Self-Honesty."

— Joshua Free, NexGen Systemology Founder

∞

SYSTEMOLOGY

mardukite.com